Environmentalism, Ethical Trade, and Commodification

This book explores the global connections between Chilean landscapes and Northern consumers embodied by the Forest Stewardship Council logo, the green seal of approval for certified sustainably produced "good wood." How do we decide what makes good forestry? What knowledges and values are expressed or silenced when "good" is defined with a market mechanism like certification? Henne's ethnographic study documents the new forms of labor and the new expectations about sustainability and responsibility that certification generates, in the context of the competing ideas about how to manage a forest—or even what a forest is—that constitute forest certification in Chile. A critical analysis of certification's practices helps understand the role of ethical trade initiatives in creating sustainable, survivable global futures.

Adam Henne is Assistant Professor of International Studies & Anthropology at the University of Wyoming.

Routledge Studies in Anthropology

1 **Student Mobility and Narrative in Europe**
The New Strangers
Elizabeth Murphy-Lejeune

2 **The Question of the Gift**
Essays across Disciplines
Edited by Mark Osteen

3 **Decolonising Indigenous Rights**
Edited by Adolfo de Oliveira

4 **Traveling Spirits**
Migrants, Markets and Mobilities
Edited by Gertrud Hüwelmeier and Kristine Krause

5 **Anthropologists, Indigenous Scholars and the Research Endeavour**
Seeking Bridges Towards Mutual Respect
Edited by Joy Hendry and Laara Fitznor

6 **Confronting Capital**
Critique and Engagement in Anthropology
Edited by Pauline Gardiner Barber, Belinda Leach and Winnie Lem

7 **Adolescent Identity**
Evolutionary, Cultural and Developmental Perspectives
Edited by Bonnie L. Hewlett

8 **The Social Life of Climate Change Models**
Anticipating Nature
Edited by Kirsten Hastrup and Martin Skrydstrup

9 **Islam, Development, and Urban Women's Reproductive Practices**
Cortney Hughes Rinker

10 **Senses and Citizenships**
Embodying Political Life
Edited by Susanna Trnka, Christine Dureau, and Julie Park

11 **Environmental Anthropology**
Future Directions
Edited by Helen Kopnina and Eleanor Shoreman-Ouimet

12 **Times of Security**
Ethnographies of Fear, Protest and the Future
Edited by Martin Holbraad and Morten Axel Pedersen

13 **Climate Change and Tradition in a Small Island State**
The Rising Tide
Peter Rudiak-Gould

14 **Anthropology and Nature**
Edited by Kirsten Hastrup

15 **Animism and the Question of Life**
Istvan Praet

16 **Anthropology in the Making**
Research in Health and Development
Laurent Vidal

17 **Negotiating Territoriality**
Spatial Dialogues Between State and Tradition
Edited by Allan Charles Dawson, Laura Zanotti and Ismael Vaccaro

18 **HIV/AIDS and the Social Consequences of Untamed Biomedicine**
Anthropological Complicities
Graham Fordham

19 **Environmentalism, Ethical Trade, and Commodification**
Technologies of Value and the Forest Stewardship Council in Chile
Adam Henne

Environmentalism, Ethical Trade, and Commodification
Technologies of Value and the Forest Stewardship Council in Chile

Adam Henne

NEW YORK AND LONDON

First published 2015
by Routledge
711 Third Avenue, New York, NY 10017

and by Routledge
2 Park Square, Milton Park, Abingdon, Oxon OX14 4RN

Routledge is an imprint of the Taylor & Francis Group, an informa business

© 2015 Taylor & Francis

The right of Adam Henne to be identified as author of this work has been asserted in accordance with sections 77 and 78 of the Copyright, Designs and Patents Act 1988.

All rights reserved. No part of this book may be reprinted or reproduced or utilised in any form or by any electronic, mechanical, or other means, now known or hereafter invented, including photocopying and recording, or in any information storage or retrieval system, without permission in writing from the publishers.

Trademark Notice: Product or corporate names may be trademarks or registered trademarks, and are used only for identification and explanation without intent to infringe.

Library of Congress Cataloging-in-Publication Data
Henne, Adam.
  Environmentalism, ethical trade, and commodification : technologies of value and the Forest Stewardship Council in Chile / by Adam Henne.
     pages cm. — (Routledge studies in anthropology ; 19)
  Includes bibliographical references and index.
  1. Sustainable forestry—Chile. 2. Forest management—
Chile. 3. Forests and forestry—Environmental aspects—Chile. 4. Forest products—Chile—Marketing. 5. Forest products industry—Chile.
6. Forest Stewardship Council. I. Title.
  SD161.H46 2015
  634.9'20983—dc23   2015002415

ISBN: 978-0-415-73041-9 (hbk)
ISBN: 978-1-315-81924-2 (ebk)

Typeset in Sabon
by ApexCoVantage, LLC

Printed and bound in the United States of America by Publishers Graphics, LLC on sustainably sourced paper.

Contents

	Preface: Knowledge and Nature	ix
1	Introduction: Good Wood	1
2	Making Wood and Making Persons	27
3	Putting Knowledge to Work	56
4	Green Lungs	82
5	Certification and the Politics of Scale	95
6	Conclusion	122
	Bibliography	135
	Index	153

Preface
Knowledge and Nature

It is about seven o'clock in the morning, and like most mornings, I am fending off a playful toddler while I try to light a fire in the woodstove. This woodstove will feature prominently in the upcoming narrative about forest certification in Chile, just as it featured prominently in my life in Chile and my self-image as an ethnographer. More on that later. At the moment, building a fire is simply the setting in which I encounter a strange piece of documentary evidence. I am crumpling up newspaper and shoving it into the stove, mostly failing to keep it away from my child, when my eye falls on an advertisement spread across the last page of the local paper. The ad shows a large full-color photograph of a local landmark, the Pedro de Valdivia Bridge that crosses the Valdivia River just a few blocks from our *cabaña*. Superimposed on this image is a group of four smiling young Chileans, wholesome and very white. "Universidad Austral," reads the caption, "Knowledge and Nature at a Real University."

"Knowledge and Nature" (*conocimiento y naturaleza*), it turns out, is the motto of the Universidad Austral de Chile, appearing in all of their advertisements, on billboards and T-shirts and letterhead. The rest of the text is easy to parse: ". . . at a Real University" is marketing hype, the kind of claim found in any advertisement. Universities in Chile's privatized higher-education system compete for students (and their money) just as the students compete for a dwindling pool of grants and scholarships.

In the context of this competition, ". . . at a Real University" is not just a claim of authenticity, but an act of boundary work. This term first appeared in the social studies of science, to describe the process of discursively separating 'science' from 'not-science' (Gieryn 1999); but it can be applied equally well to any number of social projects anxiously policing the border between 'us' and 'not-us.' As an organizing concept, boundary work will appear in a variety of configurations throughout this project. Tracking the overt strategic use of discursive tools like this one is the first of the analytic principles that I'll be using here.

But back to the newspaper ad; "Knowledge and Nature" is a little more obscure. As the university's motto, it circulates well beyond advertisements, and is made to do rather more complex representational work. A motto

x *Preface*

is a strange genre of text, laden with signification but virtually devoid of meaning. That is to say, these are two big and powerful words, but what are they supposed to be telling us about the university? Really, they only make sense in the context of being read by particular readers, who thereby impart particular meanings to them. Understanding how significant objects become meaningful through circulation and interpretation is my second tool—"knowledge" and "nature" are themselves two of the most significant and most powerful objects that will circulate through this story.

Before turning to the theoretical concepts that move my interpretations, we should stop by another significant motto that circulates widely in Chile. Printed around the edge of the 100-peso coin are the words "By Reason or By Force." The original meaning of this ominous claim is now somewhat murky.[1] But it's resonant of the threat implied by the omnipresence of the state, the "power" side of the power/knowledge dyad, and the mild, persistent paranoia of those who live in a post-dictatorship regime. In the pocket of every Chilean is a quiet reminder that reason and knowledge exist in a context defined by the exercise of more overt forms of power, by the presence of economic domination and state violence. So "By Reason or By Force" leads me to the third analytic tool that drives my discussion of certification; successive re-contextualization of knowledge in the fields of power and political economy that shape it and give it force.

But why am I here, building fires in a cabin in southern Chile? Mottos and advertisements aside, I am here to study forest certification. In the following section I will provide a fuller explanation of certification and how it fits into growing worldwide projects of green capitalism and ethical trade. For the moment, I defer to its more visible symbols: Shown here, the label marking wood or wood products that have been certified by the Forest Stewardship Council.

Figure 0.1 Forest Stewardship Council logo.

Protected by copyright and supported by an international coalition of companies and non-governmental organizations, this label both communicates and enforces the standards of sustainable forestry the FSC's stakeholders have painstakingly defined. I have dragged my family here to the Southern Hemisphere, to the land still sometimes referred to as 'the ends of the earth,' to see what this symbol can do on the ground.

NOTE

1. Many of us visiting Chile imagine the motto to be a legacy of the Pinochet dictatorship, but apparently it predates the foundation of the independent state (Loveman 2001).

1 Introduction
Good Wood

NEOLIBERAL? HARDLY!

My fieldwork on the Forest Stewardship Council in Chile had just begun when I managed to get myself invited to the annual retreat of the Association of Foresters for Native Forest (AIFBN), an advocacy group whose members have been central to the formation of the FSC here in Chile. Seated on tree stumps with the president and secretary of FSC-Chile, I was called on to introduce myself and explain my research project. In particular the president, Luis Astorga, was a little suspicious of my intentions; am I out to smear the FSC? I reassured him that it's not my approach at all, and that I was interested instead in how different kinds of knowledge, authority, and models of nature play together in building a neoliberal form of environmental governance like forest certification. I should have known better than to use the term 'neoliberal' so lightly. Astorga leaned back indignantly: "Neoliberal? Hardly! Certification is a way to include non-market values like biodiversity that economists don't even want to consider. We're not here to greenwash the timber companies—we're providing the tools to change them completely."

This was not the place to quibble. In some ways the dynamic Astorga described, using a non-state market-based form of governance to turn economic externalities into internalities, was precisely a neoliberal program. But the term has a different tone in Latin America, even fundamentally neoliberal Chile, than it does in the US; I might as well have called his organization 'imperialist.' Astorga rejected neoliberalism as a frame for the FSC because, more than assigning a price to currently unpriced values, it forces capitalism to become what it doesn't want to be. Fortunately I was able to convincingly frame my gaffe as tangential to my real interest, the procedure or techniques by which certification makes the ethical practical.

Like Fair Trade or other forms of market-based social change, FSC certification ostensibly depends on a market premium on sustainably produced wood to push producers toward more sustainable practices. This dependency implies global connections between Northern consumers, Chilean producers, and the physical landscape of Chile itself. The value-based standards that attempt to constrain those global connections are the product of political contests not visible in the wood products at the end of the commodity chain. Of the many global attempts at market-based forest governance, the FSC has made the most convincing efforts to, as they put it, "bring people together to find solutions to the problems created by bad forestry practices and to reward good forest management." That bringing-together is productive not just of timber, but of powerful ideas about sustainability, responsibility, and value.

An approach to forest protection based in resource management and sustainable use rather than logging bans and preservation, certification attempts to use the power of the market to influence change in forest management practices. In the case of the FSC, principles and criteria for sustainable forest management are defined first at the global and subsequently at the national or regional level. FSC-International accredits individuals or organizations as auditors, who may evaluate a given forest management unit. If management meets established standards, the auditor may award certification. Products from that management unit then carry a copyright-protected seal identifying them to consumers as certified sustainably grown wood.

Theoretically, at least, consumers will pay a premium for certified wood, thus providing a financial incentive for producers to practice sustainable forestry. Hence, analysts refer to forest certification and similar regimes as "non-state market driven" governance (NSMD;[1] see Cashore et al. 2004) or "soft law" (Kirton and Trebilcock 2004). I chose the Forest Stewardship Council over other certification schemes not because they are representative of the practice as a whole. Rather, they represent the ideal-type; certification 'done right' (Ozinga 2001; Counsell and Loraas 2002) with the most stringent standards and transparent, representative processes.

Although these governance projects are generally described as "values-based," the values in question warrant further consideration. In

Figure 1.1 Forest Stewardship Council logo.

practice they are the dependent outcome of a complex and often sub-rosa negotiation, shaped by power relations and instrumental constraints like time, bureaucratic forms, and the attention span of institutions. FSC certification, after all, is designed by its stakeholders; ENGOs, workers, indigenous people sit down with timber companies to hash out the terms that will define good forestry for the purposes of certification. Understanding how conflicting interests, values, and epistemologies come together in such a project means considering forest certification cosmopolitically, as an encounter across difference that ramifies in unpredictable ways. Existing dynamics may be reinscribed or transformed as differences and discourses proliferate (Stengers 2010; Tsing 2005). Institutional outcomes are often path-dependent; constrained by its infrastructure, FSC certification reinscribes certain existing power relations in Chile in ways that support a reading of it as categorically neoliberal governance. At the same time, other possibilities emerge as actors engage certification differently, according to oblique agendas. On paper FSC certification might look like a classic market intervention, turning environmental externalities into internalities. But timber companies, NGOs, activists, and consumers are not *only* rational agents in an abstract economic model; their lived encounters and experiences open up spaces for other kinds of political rationalities.

Forest certification is one of several aftershocks of the high-profile rainforest protection campaigns of the late 1980s.[2] Environmental NGOs promoting boycotts of destructive tropical timber needed a system of alternative recommendations; a way for them to help consumers find wood products they could feel good about buying. Organizations including the Rainforest Alliance and Greenpeace collaborated in devising the systems of third-party certification that eventually developed into the Forest Stewardship Council and other certification schemes.

In general, a certifying body like the FSC defines principles and criteria for sustainable forest management at the global, national, or regional level. The certifying body accredits individuals or organizations as auditors, who may evaluate a given forest management unit. If management meets established standards, the auditor may award certification. Products from that management unit then carry the seal of the certifying body, identifying them to consumers as certified sustainably grown wood. Hypothetically, at least, consumers will pay a premium for certified wood, thus providing a financial incentive for producers to practice sustainable forestry. In other words, like certified organic agriculture,[3] forest certification relies on market forces rather than government enforcement to achieve compliance. Hence, analysts refer to forest certification and similar regulatory regimes as "non-state market driven" regulation or "soft law" (Kirton and Trebilcock 2004).

Given the continuing absence of a global forest convention, environmentalists are beginning to see certification schemes as an alternative regulatory approach with real potential for saving the world's forests. Because of regulatory flexibility and the opportunity to collaborate in their design, states

4   *Introduction: Good Wood*

and timber companies are also increasingly willing to invest in certification systems (Cashore et al. 2004). Since the FSC began operations in 1994, they have certified nearly 100 million hectares in 82 countries, via a network of National Initiatives in 43 countries and counting (FSC Annual Report 2005).

Most existing analyses of forest certification systems fall into one of two categories. On the one hand, practitioners and interested professionals, from forest managers to conservation biologists, have questioned the ability of certification to achieve its intended outcomes. Biologists are skeptical that certification standards can adequately measure and manage the complex and unpredictable requirements of diverse ecosystems (Brown et al. 2001; Gullison 2003). Forest managers and timber industry researchers are concerned that certification requirements have yet to provide a price premium to offset their costs. As a result, certification standards may be in the process of becoming an unrewarded minimum standard for participation in the market that will exclude many producers (Haener and Luckert 1998; Nebel et al. 2005). With the advent of carbon market initiatives like REDD (Reducing Emissions from Deforestation and Forest Degradation), forest certification may become an even larger aspect of the global timber market.

The other set of discussions emerges from institutional policy analyses by economists and political scientists. They want to understand why different certification schemes appear in various regulatory environments, what factors influence the participation of producers or NGOs in certification projects, and how the move toward non-state regulations affects the credibility of sustainability claims (Eliot and Schlaepfer 2001; Meidinger, Eliot, and Oesten 2003; Rametsteiner 2002). The most comprehensive ongoing analyses of certification, conducted by political scientist Benjamin Cashore and his research group at the Yale Program on Forest Certification, includes an exhaustive comparative study of FSC certification in Europe and North America, taking into account existing forest regulation, NGO pressure, and market dynamics (Auld et al. 2008; Berstein and Cashore 2004; Cashore et al. 2004; Cashore et al. 2007; McDermott et al. 2009; Newsom et al. 2005). The extent to which forest certification succeeds in any given case at its tripartite goal of protecting forests, promoting social justice, and generating profits clearly depends on a complex set of variables with an extremely wide range of possible outcomes.

TECHNOLOGIES OF VALUE

Forest certification is an increasingly common form of environmental governance, and it clearly offers a variety of toeholds for analytical consideration. I find myself unsatisfied, though, by the institutional approaches analysts have taken to date. To my mind, they leave unexamined the most intriguing implications of forest certification—its cultural politics. While my

Introduction: Good Wood 5

project shares an object of study with this literature, my approach draws more from political ecology and science studies; hence, the kind of questions I ask address certification as a cultural artifact. Certification through the FSC, after all, is a system of regulation designed by its stakeholders. This means that competing actors sit down to negotiate what kinds of practices constitute 'good' forestry. What kinds of relationships are implied (or created) when they sit down like this? What kinds of meanings are generated in the process—for keywords like 'sustainability,' 'credibility,' or even 'forest'? In other words, although labeling schemes like FSC are generally described as "values-based" forms of regulation (Barham 2002; McAfee 1999), the values in question are neither transparent nor pre-determined. Rather, they are the dependent outcome of a complex and often sub-rosa negotiation, shaped by power relations and instrumental constraints like time, bureaucratic forms, and the attention span of institutions. As a result, I would argue that making sense of the label (the FSC logo™) and the values it represents (sustainable forestry) ethnographically allows me to explore the ways that certification's lines of force locate political agency and responsibility.

To help structure this approach, I would like to introduce the term 'technologies of value.' I consider FSC certification to be a technology not in the sense of machines and space-age polymers, but rather as a technology of government. Like all technologies, governmental technologies require actors to mobilize multiple forms of knowledge and combine them in more-or-less stable entities in order to achieve their ends. As Rose puts it, a technology of government consists of:

> lines of connection among a diversity of types of knowledge, forces, capacities, skills, dispositions, and types of judgment ... forms of practical knowledge, modes of perception, practices of calculation, vocabularies, types of authority, forms of judgment, architectural forms, human capacities, non-human objects and devices, inscription techniques and so forth. (1999, 52)

It is precisely this definition I have in mind when I consider the relationships between the multiple subject positions entangled by FSC-Chile, and the divergent forms of knowledge they bring to the table. Thus I am situating this project in the long lineage of anthropology that puts Foucault's concepts of governmentality and power (1980, 1991) to work in situations of cultural difference. Anthropology in the 21st century has renewed its interest in the power at work in bureaucratic forms and the generative qualities of government.[4] Studies of governmentality in the sphere of what we call "the environment" make up a small but significant subset of this literature (Darier 1996; Luke 1999; Rutherford 1994). Agrawal has insightfully documented what he terms 'environmentality'; the ways that environmental regulation, as a technology of government, generates in its subjects new ways of imagining themselves as environmental actors (2005). Geographers in particular have elaborated on technologies of environmental governance as spatial forms and instruments of political rationality (Bryant 1998; Bryant and Goodman 2004; Heynan et al. 2007; Rocheleau and Roth 2007). Because they are so heterogeneous in terms of spatial configuration and political subjectivity (Barnett et al. 2011), ethical consumption movements like forest certification or Fair Trade are particularly rich settings for such analysis.

I add to this argument the concept of 'value,' a term whose polyvalent meanings are splashed all across the domain of forest certification. In one sense, forest certification is intended to generate economic value for timber producers. Ideally, wood carrying the FSC logo brings in higher prices at the retail end of the commodity chain. This surplus value, when it makes its way to the wood producer, compensates them for the higher costs associated with more sustainable forms of production. Certification then is a technology designed to create and stabilize economic value. In another sense, forest certification is intended to represent the ethical values of consumers. The FSC logo on a given piece of wood indicates that it was produced with respect for biodiversity, indigenous territory, maintaining resources for future generations, and other values collectively understood as 'sustainable forestry.'[5] The FSC directs its material and intellectual resources towards ensuring that production practices really do measure up to the promise of the logo. In this sense, certification is a technology for enabling, or even imposing, the values of consumers. The tension or synergy between these two aspects of value constitute the explicit basis for certification's purpose and function. The members of FSC-Chile debate with each other the relative weight to be placed on each aspect, and debated with me the extent to which certification is "neoliberal," for better or for worse; but all agreed on the fundamental encounter between economic value and environmental values at the heart of the process.

The third sense of value that I propose is more subtle, and locates certification as a technology within the domain of governmentality. Through the course of the book, I will demonstrate how actors associated with FSC-Chile develop new ways of *valuing*, new tools for engaging with the market and 'the sustainable.' The distinctive contribution of 'technologies of value' to analysis of governmentality more generally is my argument that the first two aspects of a technology of value are productive of the third. That is, the active efforts of the FSC to produce economic value and promote ethical values contribute to new ideas about what it means to have value and values. Furthermore, they do so in part by eliding the connection between actor's positions and the outcome of the regulation—by naturalizing the process. Where the economic, the ecological, and the ethical come together, more than a simple regulation is at stake. A technology of value, like a technology of government in Foucault's sense, is a lever for shifting the locus of political power from one set of actors to another. While this depends on the explicit content of the regulations themselves and the resources available to enforce them, it is equally entangled with the definition of the individuals or collectives involved. In other words, the process of creating connections between economy and ethics defines the ethical subject of environmental regulation, attempts to fix an answer to the question "*Who* cares about the environment?" (Federovisky 2007). Like Foucault's governmentality, technologies of value are frameworks for "the government of one's self and of others" (Burchell et al. 1991, 2). Analyses that treat ethical consumption systems as social movements rather than patterns of consumer behavior (i.e. Barnett et al. 2011; Clarke 2008; Jaffee 2007) highlight the diversity of political rationalities that can emerge from the application of technologies of value.

Speaking of FSC certification as a technology of value allows me to avoid some of the problems in the existing research on certification. The debates that essentially ask "Does it work or doesn't it?" (i.e. Brown et al. 2001; Gullison 2003; Haener and Luckert 1998; Nebel et al. 2005, among others) propose a binary opposition that overlooks the multiplicity of definitions at work here. The institutional discussions (Cashore et al. 2004; Eliot and Schlaepfer 2001; Meidinger, Eliot, and Oesten 2003; Rametsteiner 2002) break up the binary, but accept categories like *timber producers, environmentalists*, and *the state* as uncomplicated givens. If instead I trace the 'lines of connection' (Rose 1999) that produce the FSC logo, I can bring in a whole set of conceptual tools to understand *why* actors participate in the FSC, *how* they deploy discursive and material strategies to shape its outcomes, and *what* sort of new understandings emerge from the encounter.

I discuss those tools at length in the relevant chapters; below I introduce only some essential foundations. First, though, I need to explain why the FSC experience in Chile is a particularly rich and complex example of the ways in which nature and knowledge are constructed and articulated in this new regime of governance.

GOOD WOOD IN CHILE

Most of the people I interviewed would get curious at some point: "Why here? Don't you have [Indians/forestry/environmental conflicts] in the US?" And we do, we certainly do. I admit to feeling a little put-upon by the question sometimes; earlier generations of anthropologists didn't have to defend their field sites like this. My response is symptomatic of anthropology's troubled relationship with colonial history, but in a postcolonial era of multi-sited ethnography and problem-based research, pat answers to this question no longer hold water. Fortunately, a series of ecological, economic, political, and cultural circumstances all lined up to justify my presence.

The forests of Chile are often described as some of the most impressive in the world. Most of the remaining native forests in Chile are temperate rainforests, an increasingly threatened ecotype. Chile possesses about

8   *Introduction: Good Wood*

one-quarter of the global total of temperate rainforest (Armesto et al. 1995; Wilcox 1996); if one considers only temperate rainforests that remain largely intact, that total rises to one-third (Neira et al. 2002).

Temperate rainforests are not as rich in biodiversity as their tropical cousins; however, due to millenia of geographical isolation, species native to Chilean forests show a very high degree of endemism (Aagesen 1998). Twenty-eight out of 84 plant species are endemic, as well as 11 species of mammals, 24 species of amphibians, and 13 species each of birds and fish (Neira et al. 2002, 13). In addition to their biodiversity, Chilean forests serve important ecosystemic functions; maintaining hydrological cycles and soil stability, protecting watersheds, and even acting as a carbon sink due to their exceptionally large standing biomass (Neira et al. 2002). Little wonder, then, that environmentalists have taken an interest in preserving Chilean forests. As Global Forest Watch points out:

> The global significance of Chile's forests has been recognized by multiple well-known international conservation organizations such as World Wildlife fund, Conservation International and the World Conservation Union. WWF, for example, has catalogued Chile's temperate forests as one of the top conservation-priority forest ecoregions in the Southern Hemisphere [Global 200], while CI and IUCN have identified Chile's forests as one of the 25 "hot spots" for biodiversity conservation in the world. (Neira et al. 2002, 18)

We need to understand these priorities and valorizations in the context of threat and environmental change; and indeed the Chilean temperate forests are threatened. Native forest cover is currently being lost at a rate of 120,000 to 200,000 hectares each year. Chile's Central Bank has estimated that at this rate, all native forest throughout the central region of the country could be gone within 25 years (Wilcox 1996).

As Rodrigo Catalán has pointed out, a number of factors are at work in this deforestation process. Local residents clear land for agriculture and grazing, or cut trees for firewood, and forest fires spread rapidly over pasture and degraded forest land. The greatest single factor driving the loss of native forest, however, is the expansive growth of the plantation timber industry (Catalán 1999).

It is difficult to overstate the importance of the forest industry to the Chilean economy. As of 2004, the export of Chilean wood products to the US alone totaled $600 million a year, and is expected to grow (Rohter 2004). The US, though, represents only a portion of Chile's market, as demonstrated by the share of Chile's gross domestic product in forestry, which grew from $390 million in 1978 to $2.2 billion in 1998. This growth puts the forest product industry second only to copper mining in terms of contribution to the Chilean economy.

Driven by this overwhelming profitability, the plantation sector has expanded rapidly. As of 1999, the government forestry agency calculated that plantations, overwhelmingly made up of exotic Monterey pine (*pinus radiata*), comprised more than 13.5 percent of Chile's total forests for a total of more than 2.1 million hectares (CONAF 1999). This marks a staggering increase from 15,000 hectares in 1940 and only 80,000 hectares even into the 1980s (Clapp 1995a). The spatial and financial expansion of the industry is accompanied by government- and private-funded media support, including television ads and billboards claiming Chile as "Timber Country."

Supporters of the plantation model of forestry production (Hartwig 1994), argue that growth of the plantation industry is a positive development for native forests. Plantations are seen as replacement forests, taking the pressure of harvesting off of fragile native forests, or in other words, "if tree planting and intensive management of a small area of forest can produce enough wood to meet the world's requirements, this could free large areas of natural forest from the pressures of timber harvests" (Sedjo and Botkin 1997, 15). However, as geographer Alex Clapp has demonstrated, this replacement forest model may not be accurate, at least not in Chile. Production numbers and harvesting rates show that in fact, the profitability of plantations relative to native forest, combined with weak or absent regulation, drives the conversion of native forest into new plantations. Even where new plantations are originally established on non-forested land, such as degraded pastures, the lack of incentive to conserve native forest continues to promote conversion (Clapp 1998a).

While environmentalists have concentrated their efforts on slowing the conversion of native forest, the establishment of new plantations on non-forested land is not without controversy. The majority of pine plantations are located in Chile's VIII, IX, and X Regions, an area that corresponds closely to the territory once deeded by the Spanish crown to the indigenous Mapuche. While population and land tenure has of course shifted dramatically in the 300 years since those grants, more than 200,000 Mapuche still live in rural areas of the three regions (Bengoa and Sabag 1997).

With a large and marginalized indigenous population already confined to small reservations (*reducciónes*) and facing population pressure on their lands, the arrival of a land-hungry new industry was bound to lead to conflict. Mapuche activists and community leaders argue that the expansion of the plantation sector comes at the expense of their territorial, environmental, and political rights. Many of the properties currently being converted into plantations are located within territory deeded to the Mapuche by the Spanish crown, or claimed by Mapuche cooperatives under the socialist Allende administration and ratified according to the agrarian reform laws. These lands became available to timber companies only after having been taken by force during the military dictatorship's "rollback" of the land reform (Frias 2003). Even in cases where rights to the lands themselves are

10   Introduction: Good Wood

Figure 1.2   Chile's timber regions; map modified by author from CIA World Factbook.

not in question, activists claim that the environmental impacts of industrial tree farming on neighboring properties are threatening the health and livelihoods of indigenous communities. This may take the form of pesticides, erosion, flooding, or the lack of access to firewood and wild plants (Catalán 1999; Rohter 2004; Toledo 1997). Mapuche activists protesting the presence of timber companies in their territory have also been subjected to violent repression by police and private security forces. These conflicts have led to the deaths of several activists and the lengthy imprisonment of many others under Pinochet-era anti-terrorist laws (Rohter 2004; Tobar 2003).

The legacy of the Pinochet dictatorship has marked Chilean environmental politics in a variety of ways that will continue to have repercussions as new regulatory regimes are designed.[6] The past decade's persistent land conflicts between the timber industry and Chile's indigenous Mapuche present a real conundrum for any meaningful form of environmental governance. Given the FSC's third Principle stipulating respect for indigenous territory, this ongoing and occasionally violent contest means that the creation of FSC-Chile and its standards becomes a theater for hidden (or not-so-hidden) politics. A similar potential complication lies in the definition of forests, including 'artificial forests' or plantations (Carrere and Lohman 1996). FSC-International is currently in the process of reviewing its controversial standards for the certification of timber plantations; since pine plantations overwhelmingly dominate Chile's forest sector, the outcome of Chile's negotiations may set the stage for worldwide revisions. Since the heyday of Milton Friedman and the Chicago Boys in the 1970s and 1980s, Chile has frequently been offered as a successful, even 'miraculous,' example of neoliberal reforms (Klubock 2004; Paley 2001). As forest certification, and other forms of neoliberal regulation like REDD (Reducing Emissions from Deforestation and Forest Degradation), become the norm in Chile, the outcomes and configurations that obtain there may be instructive for the emerging regulatory regime worldwide.

COMMODITY ECOLOGIES

Wood, whether certified good wood or the regular type, is a commodity. That is, it is a product produced by laborers to be sold for money in a market defined by the circulation of exchange values. Each of these terms is more loaded than the last, and for this reason I think it's important to point out now that the commodity aspect of wood does not exhaust wood's essence or potential. Wood is also, after all, the flesh of a tree, a fact with no small ecological significance. The way that the ecological, social, and symbolic capacities of trees overlap or clash with their commodity form is a key topic for this project. Nonetheless, speaking about wood and even trees as commodities opens up a whole range of subjects for discussion, and provides me with a rich philosophical background to develop my ideas about certification. The commodity is transgressive, being both stuff and thing, material and processual, rooted and transient, local and global, object and subject. Any review of the literature on commodities must begin with Marx. While neoclassical economics had defined and discussed the circulation of commodities, it was *Capital* that provided the first full-scale critical analysis of the commodity in its political context. It is with Marx's discussion of commodity circulation, and particularly commodity fetishism, that I want to begin.

Fetishism is another terribly loaded word, and it's no accident that Marx borrowed it from 18th-century colonial chroniclers and their descriptions of "primitive" religious rituals and idols (see Nancy 2001). In *Capital*, Marx is implying a kind of magic at work in the construction of capitalist society around the circulation of material goods, a black magic that alienates goods from their use-values and the worlds of their creators. Through the magic of the fetish, the real social relations between capitalists and producers (that is, exploited laborers) are rendered invisible, veiled by the representation of a material good valuable in and of itself. Money, the ultimate alienated commodity, becomes the universal token of value, while laborers, bosses, and all other social relations disappear: "It is . . . precisely this finished form of the world of commodities—the money form—which conceals the social character of private labour and the social relations between the individual workers, by making those relations appear as relations between material objects, instead of revealing them plainly" (Marx 1967, 168–9). Modern capitalism, in short, absolutely depends on obscuring the real circuits by which goods are made and circulated.

With commodity fetishism as a starting point, a project like forest certification starts to seem downright radical. Harvey enrolls these ethical trade projects in a familiar call for political resistance against the regime of commodity fetishism: "The grapes that sit upon supermarket shelves are mute; we cannot see the fingerprints of exploitation upon them or tell immediately what part of the world they are from. We can, by further inquiry, lift the veil on this geographical and social ignorance and make ourselves aware of these issues" (1990, 423). Latour (2010) describes such modernist anti-fetish projects as iconoclasm, smashing the idol of false consciousness to liberate the authentic knowledge of oppressive social relations. In this sense, forest certification and comparable "non-state market-driven" (NSMD) forms of regulation[7] represent attempts to reconfigure commodity networks to be more socially just and/or ecologically sound. A key point of debate is whether Harvey's defetishization is taking place, or if standardization and market regulation tend to co-opt movements for sustainability and social justice into frameworks that stifle their critical arguments. Some researchers have emphasized the positive potential of ethical trade programs. Raynolds (2000) describes organic agriculture and Fair Trade coffee networks as important forms of political solidarity between consumers and producers, between North and South (see also Nigh 1997). Geographer Julie Guthman, on the other hand, exhaustively studied the practices of organic agriculture in California. She found that organic certification was defined so narrowly (and politically) that it could be applied to agroecological practices from the most rigorous sustainable farm to the largest intensive monocrop (Guthman 2000). In short, the seal of organic certification carried no real information about production to the consumer, and thus served as another layer in the veil of commodity fetishism. Similarly pessimistic studies have been conducted on shortfalls in producer–consumer links in Fair Trade coffee programs (Mutersbaugh 2002; Rice 2001). Both sides of such debates, though, illustrate the limitations of commodity fetishism as a frame for understanding contemporary global economies. The consumer can't possibly know the production of a wood product in the physically embodied sense one might imagine for a Marxist worker *not* alienated from the product of his or her labor. What do any of us know about the sources and relations of production of the things we use every day—what kind of labor built my furniture, what's in this soft drink, how does unleaded gasoline get to my local Exxon station? Could we ever be expected to know these things?

The FSC logo, like a Fair Trade or certified organic label, functions as a proxy for that knowledge, enabling a consumer to make an ethically informed decision in the absence of knowledge. Even a carefully labelled, ethically produced consumer good is still a commodity, and as such the source of its value is separated from its production. One of the more theoretically sophisticated examples of the ethical trade literature (Whatmore and Thorne 1997) considers the production and certification of Fair Trade coffee from the perspective of actor-network theory. They describe the system of provision for Fair Trade coffee as a global network made up of actors and knowledge-producers of many types, each of whom must be enrolled in terms of their own interests in order to make the whole function. Therefore the whole geographic knowledge of the 'ethical' in the context of Fair Trade must be considered a 'network effect,' a contingent result of distributed efforts rather than an autonomous ideal. As I will discuss in more detail in Chapter 3, this perspective, which Whatmore and Thorne characterize as 'caring at a distance,' is an important foil for my 'technologies of value.' As a technology, it is the logo itself that matters here, literally: Good wood is not materially different in substance from non-certified wood. Nominally the conditions of its production will be different, having demonstrably met the standards established by the FSC.[8] But even to the extent that certification documents legitimately distinct productive practices, good wood is made and not grown; the goodness is a social relation and not a property of the wood itself. Instead of an unmediated truth of production and provision, certification offers an alternative set of narratives. Goodman compares narratives of this kind to Harvey's description of conventional commodity fetishism: "In fair trade, the oft-cited 'silent grapes' are veritably *shouting* their alternativeness to consumers" (2004, 893). Neither a fetish that passively veils oppressive social relations, nor a transparent indicator of ethical productive practices, the FSC must be read as a network effect and a political technology.

KNOWING THE FOREST

> So we see how a technological norm gradually reflects an idea of society and its hierarchy of values, how a decision to normalize assumes the representation of a possible whole of correlative, complementary or compensatory decisions. (Canguilhem 1991, 246)

Treating forest certification as a technology opens conceptual possibilities not limited to Foucault's analogical technologies of government, of self, and of care. Science studies as an interdisciplinary field offers analytical tools applicable not only to classically technoscientific objects of study, but to many of (post)modernity's entanglements as they present themselves in the world beyond anthropological categories of culture, locality, and discipline.[9] Much of my topical material is explicitly the technoscientific; the baseline for understanding sustainable forestry is the theory and practice of forest ecology, a classic case of what Foucault called "a dubious science."[10] Methodological concerns are equally compelling, though: How to understand the world-making practices of actors interacting across a spectrum of political positions and knowledge forms. And finally, network-based methodologies imply a specific sort of ontology as well, one that is willing to consider global wholes only as the outcomes of their particular networks. Although science studies and anthropology have a lengthy history (Hess and Layne 1992), ideas about knowledge and non-human agency borrowed from science studies have recently made a major entry into anthropological theory.[11] Of particular relevance here is Isabelle Stengers's concept of *cosmopolitics*, referring to practices and struggles across difference where "cosmos refers to the unknown constituted by these multiple divergent worlds and to the articulation of which they would eventually be capable" (Stengers 2005, 995)— what Donna Haraway often calls *worlding* (2007). Anthropologists have made particularly pointed use of this concept to reframe indigenous political action (Blaser 2010; de la Cadena 2010; Escobar 2008 and 2010), refusing easy relativist readings of indigenous spirits, gods, or earth-beings as *only* beliefs or representations. To my mind, network methodologies and political ontologies harmonize quite well with contemporary anthropological takes on global connection (i.e. Tsing 2005); with ethnographic studies of commodity circulation (see above); and with the deliberative network-building process of the FSC itself. What I hope to bring to this discussion is the chronicle of specific practices that connect international trade, consumer and environmental ethics, and the forms of knowledge (including but not limited to forestry science and conservation biology) that they harness to guide them.

The Forest Stewardship Council, in Chile as elsewhere, positions itself as a regulatory gateway for sustainable forestry through the definition of standards. Standards like those defined by the FSC are a very specific technology for the management of people and materials at a distance. Latour describes standardization as a strategy for achieving universality: "The history of technoscience is in a large part the history of the resources scattered along networks to accelerate the mobility, faithfulness, combination, and cohesion of traces that make action at a distance possible" (1987, 287). Compare this to the guiding principle of Fair Trade networks, whereby more direct producer–consumer connections allow for an ethic of "caring at a distance" (i.e. Whatmore and Thorne 1997). FSC, Fair Trade and other commodity networks play by many of the same rules as other, more classically technoscientific networks beloved by STS, but have yet to receive the same level of attention.

One significant exception is the work of Lawrence Busch and his colleagues on standardization in agricultural science and practice, which bears citing in detail:

> Through the use of product grades and standards, we subject nature and people to rites of passage. . . . The use of such standards gives rise to at least three symmetries: (1) successful completion of a rite of passage by a nonhuman is always simultaneously a statement about the goodness of both the nonhuman and the human actor; (2) by defining the good, standards and grades make nature and people more uniform, measurable, and controllable; and (3) by reducing the heterogeneity of the behavior of both people and things, standards make *both* capitalist markets and neoclassical economics possible. Thus, by transforming nonhumans and subjecting them to multiple rites of passage, we coproduce nature, society, the capitalist market, and neoclassical economics. (Busch and Tanaka 1996, 5)

I appreciate the framework that Busch lays out here, and I think that these implications of symmetry have a great deal of potential relevance to the FSC scenario in Chile. At the same time, I think the reference to a unitary "we" who coproduces nature, society, and the market loses the nuance of Latour's networks of distributed agency, and makes the process seem a little too linear and predictable. Agrawal (2005) refers to the emergence of environmental subjectivities in response to regulatory regimes as *environmentality*; hence we might look to standards and standardization, where multiple actors and their forms of knowledge connect and compete across space and cultural difference, as a site of "*distributed* environmentality."

One example of such a distributed, historicized perspective can be found in Timmermans and Berg's work on standardization of medical protocols. They describe standards as "technoscientific scripts which crystallize multiple trajectories. In the process of obtaining local universality . . . protocols feed off previous standards and practices" (Timmermans and Berg 1997, 273). In the case of the FSC in Chile, standards are built from pre-existing materials in ways that are explicit (i.e. the first criteria for certification stipulates compliance with existing forest laws) and implicit (i.e. the first draft of the standards was drawn up using Bolivian and Finnish standards as models). In other words, standards do not represent *sui generis* expressions of moral will, as the quote from Canguilhem might suggest, but rather the contingent processing and re-contextualizing of existing practice. Furthermore, this process is contested and uneven: "protocols function through the distributed work of a multitude of actors . . . in this process, protocols themselves are necessarily changed and partially reappropriated" (Timmermans and Berg 1997, 273; also see Timmermans and Epstein 2010). It's this distributed work across viewpoints that this book attempts to document, pulling together the perspectives of actors at every node of the network making up FSC certification in Chile; conservation biologists, foresters, plantation workers, activists, and others.

Theories of distributed knowledge-making, like Latour's actor-network theory of technoscience or the 'crystallized scripts' of Timmermans and Berg, often come under fire for failing to consider persistent structures of inequality like gender, race, class, and other axes of difference. Especially in contexts like that of FSC-Chile where wide power differentials influence the types of knowledge brought to the table, it is valuable to think about the constraints imposed on *where* networks can expand and *how* knowledge is distributed. For this point I defer to Bowker and Star's now-classic work on classification *Sorting Things Out* (1999). Drawing on examples from medical diagnostic technologies, archives, apartheid-era racial classification, and the training of nurses, Bowker and Star elaborate the material consequences of information infrastructure. Each stage of increasing governance delimits acceptable materials (in my case, forest management strategies) while constraining the options for each subsequent stage. Forest certification developing out of existing forest regulation, international markets, and tensions between actors is *path-dependent* in much the same way as Bowker and Star's medical or racial examples. Infrastructure in this sense is powerful because it is largely invisible, and can easily be taken for 'the way things are.' An important corollary to this fact is that, as Bowker and Star point out, "infrastructures are never transparent for everyone" (1999, 1)—the power inequalities between actors in forest certification may be reinscribed or contested at every stage of the standard-making process.

I aim to describe the making of standards, then, as a project of pulling together existing forms of knowledge across significant epistemological and political divides, producing a network or a script that is at best a contingent lash-up of perspectives. Science studies provides a useful spatial metaphor in the concept of *boundaries*. The term *boundary work*, proposed by Thomas Gieryn (1999), describes the social labor necessary to divide acceptable scientific knowledge from unacceptable, non-scientific knowledge. Gieryn notes that natural phenomena can sustain a number of competing explanations, and that sorting between those explanations is a matter of contestations of legitimacy and credibility. While these contests use the tools of scientific authority (peer review, credentials, funding allocation, etc.), the stakes and conflicts are often political at heart. Gieryn documents this dynamic using the cases of cold fusion, phrenology, Victorian engineering, and the birth of the National Science Foundation (1999). In the case of environmental policy, scientific boundary work often takes the form of trying to purify the 'scientific' by isolating it from the 'political' (Jasanoff 1987, 1990; Yearley 2001).

As the case of FSC standards will illustrate, however, the social production of boundaries tells only part of the story. Actors in science-based projects work not only to police their boundaries but to reach across them, to translate. Policy and politics demand productive conversation between scientists, politicians, activists, and some version of an informed 'general public.' Star and Griesemer define the tools used in pursuing that sort of project as *boundary objects*, structures for translation between different ways of knowing "both adaptable to different viewpoints and robust enough to maintain identity across them" (1989, 387). This is the slippage that Tsing (2005) describes as the engine of cultural change in the global era—and most significantly, it is not a logical imperative or an accident of economic globalization, but the product of concerted intellectual and political work on the part of actors across the global stage. Are FSC standards boundary objects, and what kind of meanings emerge from their translation work? The only way to find out, it seemed to me, was to put myself in the middle of it.

METHODS, VOICE, AND AUTHORSHIP

Anna Tsing advises ethnographers who trace knowledges between scales "to focus on zones of awkward engagement, where words mean something different across a divide even as people agree to speak" (Tsing 2005, xi). With shifting definitions of key terms such as 'sustainability,' 'good science,' and 'forest,' the negotiations over FSC certification in Chile represent just such a zone.

I found my way into this particular zone of awkward engagement by feeling around in a similarly awkward fashion. My Chilean fieldwork experience began in 2004, when I spent two months interviewing participants in the Chilean Native Forests Market Campaign. I describe this campaign and its relationship to the Forest Stewardship Council in more detail in Chapter 4, but when I began the project I had very little sense of what certification had to do with the campaign or with Chilean forests in general. I was interested in the relationships between indigenous Mapuche activists and white environmentalists, and the extent to which the idea of 'indigenous knowledge' shaped their approaches to forest protection. Once on the ground, however, everyone I interviewed of all ethnic backgrounds brushed off my interest in indigenous knowledge as more-or-less irrelevant. What mattered in the context of the campaign, they told me, was not something they identified as 'culture' but rather 'politics,' by which they meant the relatively overt relations of power between Mapuche and non-indigenous Chileans. As I returned to the subject and began to make the link to certification, it became clear that crisscrossing between the 'cultural' and the 'political' would require me to cast a broad net, methodologically speaking.

The body now known as FSC-Chile was, when I began my fieldwork, called the Initiativa para Certificación Forestal Independiente (ICEFI). As a membership organization, they have a list consisting of 72 organizations and individuals ostensibly participating in standard setting to some extent. While my original intention on setting out was to interview an exhaustive

Introduction: Good Wood

ple of that membership, it quickly became clear that most of those 72 slots were redundant or so marginal as to be meaningless—more than one of the organizations I contacted had forgotten about their membership in FSC altogether. Because of this, I chose to focus on those individuals and organizations that seemed most salient according to consensus of key informants. I spent my time in the field tracing the roles of these actors in knowledge production using a toolkit of qualitative ethnographic methods—semi-structured interviews, in-depth interviews, and participant-observation. Semi-structured interviews, with a standard schedule emphasizing each organization's general platform and their role in negotiations, enabled me to locate each organization in relation to the others and create comparative categories. I conducted in-depth follow-up interviews with key individuals, providing insight into the cultural values that drive participation in forest certification and the knowledge practices that give it shape. Participant-observation at an endless parade of meetings and conferences allowed me to document unspoken interactions (e.g. who speaks first at meetings, what types of opinions are publicly treated as most credible) and "ground-truth" subjects' behavior against their statements in interviews. NGOs being the bureaucratic creatures that they are, I was able to supplement my ethnographic investigation with documentary research in the archives, published work, and gray literature of the relevant organizations, documenting the perspectives and discourses reproduced therein.

Key to my understanding of the project, though, is the conceptual move from offices and boardrooms to the forests and plantations themselves. In these settings, I investigated the social and scientific practices that take place during the certification audit of forest properties. This approach provided information on how the standards are actually applied by certification assessors, and how the social context of auditing on the ground contributes to reshaping the standards themselves. My sample in this case consisted of a stratified group of producers already certified under interim ICEFI standards, according to size and type of producer (i.e. large plantation, small plantation, native forest, sawmill, chipmill). In addition, I explored a convenience sample of producers in the process of being certified during the time that I was in residence, visiting their properties alone and in the company of the auditor.

Initially, I maintained a home base in the city of Valdivia, along with my partner and the toddler I mentioned at the outset. The city's location near the coastal temperate rainforests and the expanding pine plantations made it a convenient center for trips into the field. Several important actors in the certification process are located in Valdivia, including the World Wildlife Fund, Smartwood-Chile, and faculty at the department of forestry. As part of my fieldwork funding via Fulbright, I was affiliated with the Centro de Estudios Ambientales at Universidad Austral de Chile (UACh), which was an enormous aid in terms of making contacts and gaining access to resources for every phase of this project.

After six months in Valdivia, I shifted my base of operations to Santiago. Chileans will sometimes say that all roads lead to Santiago, and mine was no exception. Santiago holds the offices of every major timber company, environmental organization, and labor union; and anyone who doesn't have an office there has to come there on a regular basis for any kind of meeting. In Santiago I set up a sort of office (see below) within one of the FSC affiliates' offices. This placed me within the formal space of the FSC, officially acceptable to all parties without signaling a commitment to any one sector. I interviewed members of FSC-Chile's three chambers in Santiago, sat in on meetings of the FSC-Chile general board and several subcommittees, attended conferences on forest policy and the new Ley de Bosque Nativo [Native Forest Law], and photocopied every document I could get my hands on. In the meantime I continued to travel to visit plantations and forest stands in the south, where I spent time with plantation workers and small proprietors of native forests. I made a bit of a pest of myself, while at the same time trying to make clear my support for the efforts of FSC-Chile to create a new kind of relationship between environmentalists, labor, and timber companies.

One byproduct of my fieldwork, largely unrelated to this thesis, is the result of volunteer work with CODEFF,[12] one of ICEFI's member organizations. In collaboration with the director of CODEFF's forest program and Mapuche forester Pablo Hauiquilao (see Chapter 3), I helped prepare a manual of forestry laws for an audience of small forest proprietors. The Chilean legislative dialect is dense, and forest laws are framed for the most part to deal with the issues affecting industrial timber production. Our task was to excerpt the legal documents and highlight the relevant portions in a language that less-formally educated forest proprietors could easily interpret. The end result, still in process as I write this introduction, is a booklet to be distributed to communities and individual proprietors at workshops, municipal meetings, and agricultural extension events. This collaboration, while not producing numbers or narratives bearing directly on my thesis, helped to establish the relationship that I shared with my peers and counterparts in the Chilean forest sector, and to define a mutual space of knowledge production. That is, together we put ideas onto paper and did our best to get them into circulation, to effect some changes in the way that forest and conservation policy works in Chile.

When I make this argument I am thinking partly of basic good fieldwork practice, establishing rapport with one's informants and all that, but also about examples of researcher/researched relationships in contemporary ethnography. Researchers studying politicized actors or social movements face a particular quandary in terms of their own position and responsibilities for interpretation. Many anthropologists adopt a stance of outright advocacy (i.e. Hale 2006), and in many cases the moral argument for this approach is very strong. The position of pure advocacy becomes more complicated, though, in direct proportion to the complexity of the situation in question (see e.g. Brosius 1999b). In the case of the FSC, it would be difficult for me to single out an actor I could consider 'the good guy' to support against all others. I find it far more appropriate in this case to consider the *project* of certification as a whole to be the problematic, the focus of concern, and to consider my relation to its various players from the inside. All of them are potentially agents for positive change, and all run the risks of re-inscribing the relations of domination and destruction that the FSC ostensibly opposes. That being the case, in the field I found myself frequently referring to Graeber's (2004) comments on the relationship of the intellectual to the political as 'object' of study. That is:

> When one carries out an ethnography, one observes what people do, and then tries to tease out the hidden symbolic, moral, or pragmatic logics that underlie their actions; one tries to get at the way people's habits and actions make sense in ways that they are not themselves completely aware of. One obvious role for a radical intellectual is to do precisely that: to look at those who are creating viable alternatives, try to figure out what might be the larger implications of what they are (already) doing, and then offer those ideas back, not as prescriptions, but as contributions, possibilities—as gifts.[13]

This leads me to talk a little about myself, what brought me to this research topic, and what kind of global links connect me to the subject(s) in particular ways. I grew up in rural northern New Hampshire, which means that I have lived among forests for as long as I can remember. But as my research aims to show, a forest is never just one thing. As a child, I spent as much time as I could hiking, camping, canoeing, or berry-picking. For me, the forest represented a peaceful sanctuary and a home for wildlife. On the other hand, those same trees provided a blue-collar livelihood for the entire region; everyone I knew came from a family of loggers or mill workers. The pulp and timber industry had transformed New Hampshire's social and natural landscapes in complicated ways, a reality that was never far from our attention. A humid day would bring reeking, sulfurous clouds down from the stacks of the nearby pulp mill. "Smells like dinner on the table," was the obligatory joke; but it was not a joke. The tension that I felt between these two ways of experiencing forests, two conflicting sets of values, was in fact a very productive one. I was learning from an early age that this thing we call "nature" is always more than it appears.

So to study contestations over how to define a forest comes to me pretty naturally. The personal becomes the academic in this account in more current terms as well—coming to the field as a member of a family puts me into a very different position vis-à-vis my research than the rugged lone ethnographers of anthropological tradition. Not that anthropologists have never traveled in family groups, quite the contrary; numerous examples can be found describing the role of the wife and/or child (rarely the husband) as passport into the rural community, the family providing a recognizable social entity for the indigenous interlocutor, and so forth (Isbell 1985; Cupples and Kindon 2003; Gottlieb et al. 1998; Starrs et al. 2001). My family's situation was a little different, though. Because most of my fieldwork took place in educated, professional, somewhat urban social spaces, my family was not the icebreaker that earlier ethnographers have found theirs to be. My wife was involved with academic projects of her own only handicapped by our expatriate lifestyle; my daughter charmed the neighbors but did little to bring me closer to the foresters and conservationists I was interviewing. On the other hand, being responsible for a family entering unfamiliar circumstances put me in a position to consider sustainability and consumption with an immediacy I might not have otherwise. As I describe in detail in Chapter 4, the firewood that I burned in our little cabin was a constant smoky reminder of the materiality of the consumption and governance chains that I was studying. Lutz and Collins (1993) and Haraway (1989) tell us that critique is best aimed at practices in which the scholar is personally implicated. I find myself thoroughly tied into the certification networks that link the US to Chilean forest landscapes: As a native of timber country; as an environmentalist; as a parent concerned about the world my child grows into; as a guilty liberal agonizing over the political content of all my consumer choices (also see West 2005).

All this to say that the relationships implied by the terms 'research' and 'the field' are contingent and unpredictable. Thus, I consider it important to end my thoughts on methodology with some comments on the transnational trajectory of Chilean forest politics, and how they came to intersect with my own academic path. In 2003, I had no sooner decided that the Mapuche struggle for environmental justice might be an interesting focus for my doctoral research, than a student environmental group announced that a representative of the Mapuche was scheduled to appear at our university. This was a happy coincidence, of course, but it also says something about the nature of *the field*, that mysterious place that we anthropologists go to learn about people. Traditionally, the field is far away, in the developing world, with clear signifiers of rurality and the marginal. The fact that my potential research subject was about to arrive in my town in North America to discuss the very issues I intended to study suggests that, in fact, the field penetrates everywhere. Lyn Stephen's work with politicized *campesinos* in southern Mexico investigates the blurring of the lines between the subject in "the field" and the back-and-forth presence of the researcher; she says "I realized that who I was, what I know, whom I learned it from, whom I tell it to, and how I tell it is profoundly political. [My subjects] and I were both in the same field" (2002, 14). So I thought of the upcoming visit of this Mapuche representative not only as a good sign for the future of my research, but also as a lesson in contemporary political geographies.

The fact that my field of research is so heterogeneous, in terms of space and in terms of political power, underscores the complexity of my role as researcher. If I am interviewing environmentalists from three countries, members of a transnational indigenous movement, forestry scientists, timber retailers, and government officials, which ones are "my people"; which part is "studying up"? As I pick my way through this tangled mess, my own identities—as a white man, an environmentalist, an advocate of indigenous rights, a would-be academic—are inescapably drawn into the web. I recently received an e-mail from an indigenous rights mailing list asking me to donate money to buy a new laptop computer for a Mapuche activist; the website this activist produces provided my first introduction to the Mapuche and their struggles. I am now "friends" with that same activist on Facebook; the networks through which I connect to colleagues, students, and friends are not distinct from those that link me to "the field." So the complexities of the field have incorporated me thoroughly; we couldn't call it field*work* if it were simple.

Articulating that complexity in a meaningful and readable form is hardly any simpler. As I begin to wrap up this introduction, I'd like to take some time to clarify my understanding of authorship, my stylistic method—my voice. You, the reader, have probably already noticed my relatively informal use of the first person and anecdotal style. To some extent, of course, this is a matter of my personal preference, but I think it also reflects larger concerns about methodology and representation (Clifford and Marcus 1986; James et al. 1997; Naples 2003). As this document goes on, I will also make some effort to break up the stability of this chatty "I" that I am presenting by bringing in big chunks of other people's words. I rely on extended excerpts from institutional documents and other texts produced by those whom I study. Placing their language next to and against my own is intended to highlight the contested quality of the definitions in question. I do not come to this project with a stable definition of, for example, 'forest' or 'sustainable' or 'socially just,' and neither does anyone in the FSC (whether they acknowledge that or not). By juxtaposing those texts I hope to challenge the reader to conduct some of their own analyses.[14] By the same token, I include for example a largely unedited transcript of an interview with forester and indigenous activist Pablo Huaiquilao, and some childhood reminiscences from an important environmental activist. These texts serve multiple analytical purposes, as I discuss in the relevant chapters; I mention them here to demonstrate my efforts to present an authorial voice as complex and multi-articulated as the subject matter at hand.

MAKING GOOD WOOD

The book begins by looking at forest certification as a technology of environmental governmentality, a technology of value. I recount the particular history of the FSC in Chile by tracing the subject positions that define themselves in relation to certification. That accomplished, more or less, I then use tools on loan from studies of technoscience to dissect the tendons that bind those actors together in complex combinations. These tasks are linked by a brief intermission, an interlude or narrative detour that examines a local system for certifying firewood and contrasts that context with FSC's global reach.

Chapter 2 expands on the themes of commodities and culture and links them to the circuits of knowledge that define the Forest Stewardship Council. Here I introduce the *dramatis personae*, the interpellated subject positions that make up forest certification in Chile. I begin with '*the consumer*,' the elusive figure behind all ethical trade projects, rarely speaking but often spoken for. I define the role of consumers in the FSC framework in question, emphasizing the particular histories of consuming Chilean timber and the politics that shape it. Intriguingly, it is *not* retail-end consumer demand that provides the price premium, or rent, that drives social change in forest certification. That motor is located at a different point in the network—'*the environmentalist*.' Here I recount the emergence of the FSC, on the world stage and in Chile, and describe the relationship between environmentalists' public interventions and the process of standard-making. I also introduce '*timber producers*,' an unfortunately vague term. And yet, for the Forest Stewardship Council, producers from the largest multinational to the smallest peasant woodcutter are gerrymandered into the 'economic chamber.'[15] I discuss the tensions of this awkward grouping and the social projects that brought it together, while developing the narrative of Chile's timber economy and the part played by forest certification. The chapter ends with a discussion of '*the Mapuche*,' the persistent other of Chilean forest politics. FSC-Chile attempts to represent the interests of such populations; the extent to which they succeed, and the discourses that they generate, place them in an ambivalent position in terms of indigenous politics.

Chapter 3, "Putting Knowledge to Work," deals with workers and auditors and the situated nature of their knowledge, how that relates to standards and the knowledge used (or implied) in their writing. Here I use my ethnographic encounters with plantation workers, auditors, and others close to the trees in order to explore some unmapped spaces between two bodies of literature—STS and the anthropology of indigenous knowledge. Perhaps not surprisingly, the manual laborers in the Chilean forest sector are not trained in the forestry science and conservation biology that contributed to FSC's standards. The understandings that they develop about what standards mean and what they are for are distinct and suggestive. A large portion of this chapter was written in conversation with Pablo Huaiquilao, FSC-Chile's sole indigenous representative and a university-trained forestry engineer, with a particularly insightful take on indigenous knowledge in a post-colonial context.

Chapter 4, "Green Lungs," serves as a sort of intermission. I place my consideration of FSC-Chile against a similar endeavor taking place on a much smaller scale; a system for certifying the production and sale of

24 *Introduction: Good Wood*

firewood. My narrative about firewood is intended to bring up some issues of my own presence in the field with a body and a family, as well as complications revolving around environmental regulation on the scale of the local and the global.

In Chapter 5 I turn to the politics of scale and their expression throughout the standard-making process. The spatial metaphors of science studies, boundary work, and boundary objects, lend some shape to my discussion of arguments within the FSC. Scalar issues are particularly pertinent in considering the appearance and behavior of three interesting characters in the FSC drama; the neighboring community, the small proprietor, and the state. As an ostensibly voluntary, free-market form of regulation, FSC certification has a particularly ambivalent relation to the state. Drawing on the documents and testimony of those involved and *not* involved in FSC-Chile, I articulate some of the problematic aspects of this dynamic. While the sciences behind certification standards are concerned with spatial and temporal scale in certain explicit fashions, the project of defining and enforcing science-based standards for international trade introduces some very productive complications.

Finally, in Chapter 6 I attempt to tie these various threads back together. I have argued that standards and certification are particularly good objects for cultural study because they bring together in one site so many fields of contestation: Techno-science and international trade; indigenous and environmental movements; consumers and ethical practice. An ethnographic perspective illustrates how the FSC and its knowledge practices work together to produce new subjectivities while perhaps re-inscribing existing structures of inequality. Analytically, I treat these outcomes as technological artifacts, in the sense of Foucault's governmental technologies or what I am here calling technologies of value. Certification's technologies of value, like all new technologies, "both presupposes and generates new forms of human labor" (Suchman 2007, 221; also see Cowan 1987). What does it mean for consumers, producers, political actors, and landscapes themselves, that environmental regulation is increasingly turning toward these technologies, toward political labor-saving devices and all their unforeseen consequences? It is my hope that this study will, at the least, raise some useful questions about the role of forest certification and other ethical trade initiatives in creating sustainable, survivable global futures.

NOTES

1. The abbreviation NSMD is pronounced "nismoid," according to the analyst who coined it (Benjamin Cashore, personal communication, March 20, 2011).
2. I address the history of the FSC and its Chilean manifestation in particular in more depth in Chapters 2 and 4. For more discussion of other institutional fallouts of rainforest mobilizations, see Braun (2002) and Brosius (1999).

3. The parallel with organic agriculture is one that I will be making throughout this work—see Chapter 2 for more detail. The analogy is of course imperfect, but the two systems share many of the same practices and challenges from both political economic and cultural perspectives (Guthman 2004).
4. i.e. Collier et al. (2004); Lakoff and Collier (2004); Rabinow (2003); Rabinow and Rose (2006); Strathern (2000); Miller and Rose (2008); Maurer (2005); Ong and Collier (2005); and conversations on an "Anthropology of the Contemporary" taking place at http://anthropos-lab.net, among many others.
5. See Raison et al. (2001) for an overview of the 'wicked problems' that lie behind defining those values.
6. For an overview of the legacy of the dictatorship in environmental politics, see Carruthers (2001). I discuss this history in relation to forest certification in more depth in Chapter 5.
7. Among these I include organic food and fiber (Allen and Kovacs 2000; Guthman 2004), Fair Trade coffee and fruits (Mutersbaugh 2002; Raynolds 2000; Taylor 2005), and sweatshop-free clothing (Klein 1990).
8. That may not be the case, as some Chilean producers argued, and a certified wood product may just be one where someone's done the paperwork (also see Newsome et al. 2005; Guthman 2004).
9. On the anthropology blog Savage Minds, regular correspondent Chris Kelty provocatively claimed "science studies *is* anthropology." His argument was not a topical one, that anthropology should properly study technoscience, but a conceptual call. Kelty argues that what lies in common between science studies and a truly contemporary anthropology "is a concern for finding the emergent and for inserting oneself inside changing practices that have a rich technical vocabulary and process. What drives all of them is the question of cosmopolitics—the collective, progressive creation of a common world" (Kelty 2007).
10. This is not to call the findings of conservation biology, forest ecology, forest agronomy, etc. into question on their own merits. Rather, for Foucault, a 'dubious science' was one whose political commitments were noticeably close to the surface—in his example, therapeutic psychiatry (1980, 109).
11. Examples include the Multi-Species Salon that has become a feature of American Anthropological Association meetings, the "Naturecultures" theme of the 2010 Society for Cultural Anthropology meeting that featured Donna Haraway as keynote speaker, and a growing body of publications that exceed or problematize the human in anthropology (see Helmreich and Kirskey 2011 for an overview).
12. Comité Nacional Pro Defensa de la Flora y Fauna [National Committee for the Defense of Flora and Fauna], the oldest conservation organization in Chile; see www.codeff.cl.
13. That is, a gift to activists rather than a contribution to policy, given that policy presumes action by the state and obscures the activist imagination—see Graeber's 'tiny manifesto' on policy: "The notion of 'policy' presumes a state or governing apparatus which imposes its will on others. 'Policy' is the negation of politics; policy is by definition something concocted by some form of elite, which presumes it knows better than others how their affairs are to be conducted. By participating in policy debates the very best one can achieve is to limit the damage, since the very premise is inimical to the idea of people managing their own affairs" (Graeber 2004, 9).
14. See also Fortun's *Advocacy After Bhopal* (2000), in which the author includes large portions of interview transcripts, organizational documents, and advocacy texts produced in collaboration with her NGO interlocutors. The point

of this expository style being to hold multiple voices in tension and allow for a more nuanced understanding of the phenomenon and relationships in question.

15. See Chapter 2 for some discussion of FSC's tricameral governing body, made up of the Economic, Social, and Ecological Chambers. Thanks to anthropologist Eric Worby for a helpful conversation in which he referred to this configuration as the 'Talcott Parsons Chamber of Horrors,' in reference perhaps to the eminent sociologist's division of social action into various functional subsystems (Parsons 1937).

2 Making Wood and Making Persons

In the previous chapter I described FSC certification as a technology of value, requiring the mobilization of multiple forms of knowledge. For this purpose it depends on global connections between Northern consumers, Chilean producers, and the forest landscape of Chile itself. Understanding those connections requires close attention to their particular histories and materialities. In this chapter I consider the context in which FSC-Chile emerged, in terms of what Anna Tsing has called 'friction': "emergent cultural forms—including forest destruction and environmental advocacy—are persistent but unpredictable effects of global encounters across difference" (2005, 3). The persistence of difference presents a challenge to environmental regulators, who tend to think of their earthly object as universal. The prescriptions for managing difference, though, defer all too often to static models of class and identity. Participatory or stakeholder-based forms of generating policy (Gregory 2000; Poncelet 2004; Joss and Berlucci 2002) suggest a limited form of representation—if we bring one of everything to the table, we'll have all the pieces for democratic action. Even in acknowledging direct conflicts, conservationists and indigenous advocates alike have tended to reify existing social categories in terms of their respective interests (Chapin 2004; Igoe 2002; Oates 1999; Redford and Stearman 1993). I fear that such general arguments based on presumed 'interests' neglect particular histories, and lead to visions of environmental politics that may be static or predetermined.

In this chapter I lay the groundwork for my discussion of FSC-Chile in terms of the literature on governmentality and subject formation. In particular, I am interested in the way that commodity systems that are global in scope generate subject positions in particular places. This global reach is what makes certification a particularly complex form of governance; I find that I can't limit my discussion to regulations or their enforcement, but must engage the commodities themselves and the collectivities and subjectivities they produce as they circulate from forest to pulp mill to Office Depot. Understanding this commodity chain, like any other global social fact, requires an ethnographic encounter with friction. Thus, as Latour might say (1993), even the longest globe-spanning network is local at every point along the way. In the case of certified wood, that peripatetic locality is embodied by paperwork; the assemblage of clipboards, invoices, barcodes, tracking numbers, and receipts known as the chain of custody:

> The FSC label can be used only on products where the chain of custody has been audited (and monitored annually). If there are several stages of processing in different plants or even different countries, each stage must be audited to ensure that the certified product genuinely originates from a specific certified forest. Any FSC-labeled product will have a chain of custody certificate number on the label and this can be used to trace the product in the event of a question arising. Products made from timber from multiple sources, such as paper and chip products, can be labeled indicating the percentage of the material that is from FSC-certified forests, provided a certain minimum percentage is certified. In these cases the chain of custody audit includes checks on the percentages of material from different sources. Keeping track of the flows of timber can be carried out in several ways, depending on the value of the product and the risk of *contamination* from other sources. These include physically marking and *segregating* the wood, bar-coding individual logs and using waybills and shipping documents to track species and volumes. (Ozinga 2001, 27, emphasis added)

The chain of custody both documents and defines a network of global connection. There is a moral imaginary at work in the chain of custody certification, with its multiple audits to prevent contamination and enable segregation. I have already compared this network of audits to organic food, similarly vetted bureaucratically to prevent the introduction of contaminants. Like organic certification, the labor in an FSC chain of custody seeks to signal to the consumer the successful segregation of materials (see Guthman 2002, 2004). Compare this to the moral and political imaginary embedded in Fair Trade certification, wherein the labor is directed towards educating the consumer about the social relations of production while transforming the networks by which the commodities are provided (Goodman 2004; Whatmore and Thorne 1997; although see also MacQueen et al. 2006). In both very different cases, the manipulation of objects (wood, vegetables, coffee) serves to materialize various ideological commitments—as Busch puts it, as "ways of defining a moral economy, of defining what (who) is good and what is bad, of disciplining those people and things that do not conform to the accepted definitions of good and bad" (Busch 2000, 274). Forest certification uses standards to shape a moral economy for sustainable forestry, and by circulating definitions of the good and the bad, contributes to shaping the individuals and collectivities that take part in it.

FORESTS AND SUBJECTS

Why *is* the production of subjects the best way to talk about the forest and how we choose to manage it? I locate this strategy in a long line of arguments regarding the social construction of nature and landscapes:[1] The common theme in this otherwise wide-ranging body of literature is the poststructural principle that the identity of non-human actors is contingent, constituted in specific historical moments. If, as Haraway says, "nature cannot pre-exist its construction" (1992, 296), then what counts as nature, what the forest can be, cannot be cleanly divided from its representations, from cultural practice and politics. Tracing the subject positions of parties to forest conflicts, then, is to locate the polyvalent entity called 'the forest' by triangulation.

Thus far I have maintained a close relationship between *certification*, the process of awarding a symbol of recognition to a producer or vendor, and *standards*, the set of rules and definitions that govern certified production. A similar conflation appears in the work of sociologist Laurence Busch on agricultural standards. For Busch, the regulatory structures of grading and standardization represent a scaling up of the 'moral economy' theses of EP Thompson (1971) or James Scott (1977). The focus on the moral content of such apparently dull legalisms as agricultural standards leads Busch to trace their practice down to the essential questions: "Who shall benefit? Who shall lose? How shall one act? What shall one do? What should the balance be between global and local production? How should markets be structured so as to insure equity? Can we develop standards that encourage more sustainable agricultural systems?" (Busch 2000, 282). What leaps out of this series of deep questions is the tension, even when discussing equity in the global and the local, between the hypothetical "one" who shall benefit or lose, and the "we" who develop standards. Even while attending to standards as the tool that produces a political subjectivity, this perspective seems to neglect the way standards work in process. I imagine this is a deliberate strategy, as black-boxing the production of standards allows for a very powerful normative argument about the outcomes that standards produce. The legacy of the moral economy and Scott's politics of legibility (Scott 1998) is clear in this claim; what is lacking is consideration of how standards are made and what kinds of politics are conjured upstream, before their segregating authority is applied. In the breakdown of Chilean subjectivities that follows, I hope to take that look upstream.

Busch's moral economy thesis developed in the context of US agriculture; an alternative perspective might be found in considering the politics of standardization in less stable settings. The European Union has been a powerful force for standardization on the world economic scene; as individual nations enroll in its projects for harmonization, novel political scenarios emerge from the encounter. Elizabeth Dunn conducted ethnographic research on

Poland's entry into the EU, and the way that producers and consumers in the Polish meat industry experienced those changes. The standards in Dunn's case studies were essentially imposed by external authorities, inasmuch as the EU requires all member nations to produce according to the same standards. Significantly, though, Polish meat producers responded to the suggestion of new agricultural standards in a variety of ways. For many, the new standards and their probable impacts on Polish agriculture were sufficient reason to oppose Poland's entry into the EU altogether. Once that path appeared inevitable, smaller producers began to opt out of the international markets to be governed by the EU, leading to the re-emergence of an informal market for unstandardized meat products not unlike that which existed under socialism. And even in those capital-intensive contexts where the standards took hold, their application was uneven; pig farmers, meatpackers, and vendors approached the legalistic requirements of standards with many of the under-the-table strategies they had developed under socialist governments. Small agricultural producers continue to agitate within Poland for substantive change in terms of Poland's role in the EU and the enforcement and interpretation of standards—a novel role for these actors that is directly related to their encounter with standardization in action (Dunn 2002, 2005).

A similarly broad historical perspective can be found in one of the major recent works to consider the emergence of political subjectivities in the context of environmental regulations. Arun Agrawal (2005) argues that policy initiatives regarding landscapes are also technologies of government capable of producing subjects, meaning people subject (or subjecting themselves) to that form of government(ality). His *subject* matter is rural Indian community forestry projects, whose engagement he traces across decades of documentary records and ethnographic study. In Agrawal's account, previously hostile villagers became over time supporters of community forest conservation, and interests previously perceived as economic came to be read in the new category of 'environmental.' Ascribing interests and practices to categories in this fashion is an important act of disciplinary power, and constrains subsequent attempts to think and act in the same sphere (Bowker and Star 1999). Agrawal's 'environmentality' bears a lot of similarity to the conceptual frame that I propose to call 'technologies of value.' Like Foucault before him, Agrawal considers the way that actors re-imagine themselves in response to the incitements of laws and regulations. Environmental regulations in particular present farmers, loggers, forestry agents, and other land managers with the opportunity to articulate discourses about authority and legitimacy with moral arguments about the order of nature. Technologies of value, like the systems of environmentality that Agrawal describes, generate power and discursive authority through putting the various actors mentioned above into combination with each other. The logics of *value* and *values* in forest certification generate connections between scales; here I aim to apply that lens to the particular collectivities, assemblages, imbroglios, or social ecologies that they call into being.

Describing certification as a technology of value requires us to think in terms of producing subjects; that is, to consider the ways that actors in forest politics interpellate with the discourse of regulation through certification. Interpellation I take from Althusser (1971), in reference to the way a subject is constituted in the moment of recognizing his/her/itself in a discursive act. The famous analogy for an authoritarian discourse is the policeman, calling out "Hey, you!" in a public place. In Haraway's interpretation, "If I turn my head, I am *a subject in* that discourse of law and order; and so I am *subject to* a powerful formation. *How* I misrecognize myself speaks volumes about both the unequal positioning of subjects in discourse and about different worlds that might have a chance to exist" (1997, 208–9). The FSC is unique among certification programs in its commitment to bringing all stakeholders to the table. Its international and regional governing bodies are tricameral, made up of the Environmental, Economic, and Social Chambers (Forest Stewardship Council 1994). Each chamber is made up of voting members assigned to one and only one such position—timber producers to the Economic chamber, conservationists to the Environmental chamber, and labor or indigenous representatives to the Social chamber. This trichotomy is neatly summarized in the FSC mission statement seen here:

Environmentally Appropriate

Environmentally appropriate forest management ensures that the harvest of timber and non-timber products maintains the forest's biodiversity, productivity, and ecological processes.

Socially Beneficial

Socially beneficial forest management helps both local people and society at large to enjoy long term benefits and also provides strong incentives to local people to sustain the forest resources and adhere to long-term management plans.

Economically Viable

Economically viable forest management means that forest operations are structured and managed so as to be sufficiently profitable, without generating financial profit at the expense of the forest resource, the ecosystem, or affected communities. The tension between the need to generate adequate financial returns and the principles of responsible forest operations can be reduced through efforts to market the full range of forest products and services for their best value.

This governmental frame is a bit of boundary work that provides good material for a sort of analysis familiar to the critical anthropologist. One of the primary concerns of this book is to figure out how the FSC governing boards propose to cleanly separate the 'environmental' from the 'social,' the 'social' from the 'economic.' I am reminded of Bruno Latour's recent book (2005) on this residual category, the 'social'; that which remains when the modern critique attempts to disaggregate the biological, the economic, the political, and the psychological. I also think about Bowker and Star's (1999) work on sorting knowledge into categories. In this case, the consequence of classification is a matter of governance. Each chamber is made up of voting members (with weighting to "achieve a balance between Northern and Southern perspectives within each chamber"), with decisions of the General Assembly requiring a two-thirds majority (Forest Stewardship Council 1994). In other words, these awkward cultural categories have been reified into voting blocs.[2]

So, who looks up when forest certification says 'hey you'?

THE CONSUMER

> I don't want to manufacture anything to be bought or sold, or buy anything sold or manufactured, or buy, sell, or manufacture anything. . . .
>
> Lloyd Dobler, *Say Anything* (1989)

First, a small tangent about consumers and the FSC. If you visit the para-academic search engine Google Scholar™, and search for scholarly publications using the string "Forest Stewardship Council," you turn up more than 10,000 hits: Primarily books, from sociology to particle physics to cookbooks. This is not, as I initially feared, because every academic is writing about the FSC. No, what the search engine found again and again was variations on this:

> Copyright (c) 19xx. PRINTED IN THE UNITED STATES OF AMERICA. Cornell University Press strives to use environmentally responsible suppliers and materials to the fullest extent possible in the publishing of its books . . . Books that bear the logo of the FSC (Forest Stewardship Council) use paper taken from forests that have been inspected and certified as meeting the highest standards for environmental and social responsibility.

Besides being a minor obstacle in my online research, this little finding is an example of the unexpected ways that global connections are played out in the consumer domain. I set out to consume knowledge *about* the FSC, by searching for it on an online database; I found myself consuming the FSC's imprimatur on the paper the very books are printed on. Surprise is a common denominator when we encounter commodity chains, as I have noted already; the fetishized quality of modern systems of provision makes every material discovery new. That sense of surprise, and its political implications, is what being an ethical or 'conscious' consumer is all about.

I begin my catalog of subject positions with the consumer in order to paint a picture of the global web of commerce that connects (sustainably) growing trees with distributors and vendors of branded sustainable products. At the far end of the commodity chain lies the consumer, whose enlightened desire for something called 'sustainably grown wood products' is ostensibly the driveshaft that pulls each link of the chain into place. Seen from the consumer end, this configuration of power may appear self-evident; of course the decisions we make about what to buy are important in the grand scheme of things. But as a larger schematic of how the global political economy functions, a consumer-based model represents a significant departure from classic perspectives. Theoretically, it seems significant that voices from cultural anthropologists (Miller 1998) to Marxists (Hardt and Negri 2000) to environmental activists (Hawken et al. 2000) have shifted their foci simultaneously. They no longer look to the working-class man as the privileged agent for political change, but rather the middle-class white woman; not the union organizer, but the shopper.

As a theoretical move, this perspective reflects the growth of a conscious consumer movement of which certified wood products are just a small part, along with organic foods and fibers, fairly traded coffees, hybrid autos, and sweatshop-free clothing. Taken as a whole, this form of consumerism encompasses concerns about health, labor rights, North–South inequity, and the spectrum of issues we know as "the environment." The impact of conscious consumerism in immediate material terms is nearly impossible to quantify, whether we mean its impact on workers, consumer health, or the environment.[3] More evident is the transformation that has taken place in the way we, as consumers, imagine ourselves a part of global economies and political change.

It's important to note what this project is and is not. I did not conduct market research on FSC products, nor did I spend significant time doing an ethnography of consumers or consumption, in Chile or elsewhere. My comments on the construction of consumer narratives are a response to the projections of hypothetical consumers on the part of other actors in the FSC ecumene; what Lockie (2002) calls 'mobilizing the invisible mouth.' Understanding consumption from the point of view of producers and other parties interested in the production side of the spectrum is a critical step in understanding ethical consumption as a distributed project, a network effect. Why are environmentalists and timber producers interested in the values of a particular set of potential consumers? And how does the identification of such a character, the 'ethical consumer,' contribute to the cultural project that is certification? When I've discussed my research with colleagues and friends, everyone has a personal anecdote or argument to make regarding these initiatives to leverage green purchasing power. Often the conversation tends towards a lament—how difficult it is for the conscientious shopper to know what's the right thing to buy, what the real impact of their decisions is, and on what scale to even consider it all. These are the concerns that led me to begin imagining certification as a technology of value, as a labor-saving device for this ethical work we do. They are also evidence, to my mind, of a political subjectivity forming around the identity as conscientious consumer. To the extent that these internal struggles shape our identities and behaviors, we are interpellating with the discourse of conscientious consumerism.

For most consumers, the Forest Stewardship Council and certified wood in general play a small role compared to the increasingly ubiquitous organic vegetables (Guthman 2004; Barham 2002; Taylor 2005). In terms of landscapes and global connections, though, the potential impact of FSC's enterprise is quite large. In their most recent public prospectus, the FSC-US attempts to define its project precisely in terms of consumer connections and global reach:

> The U.S. marketplace for wood and paper is the largest in the world. Our consumption links back to everything from South African pulp to Chilean plywood. We buy furniture made in China made of wood from Laos, and we receive millions of catalogues made from Canadian Boreal Forest fiber and from Brazilian plantations. A flooring purchase in San Francisco can link to logging in New Zealand just as much as a paper bag in a Chicago grocery store links to a pulp operation in Arkansas. Shaping how these markets ultimately accept or reject certain sources of wood has global impacts. FSC-US, while a small non-profit organization, has a large role to play in the destiny of the world's forests. (FSC-US 2006 [prospectus])

Intriguingly, while this prospectus contained enthusiastic coverage of the FSC's development of new markets for certified paper and building products, the only hard numbers it cited referred to the numbers of certificates awarded. That is, even while speaking about its outreach to consumers, FSC-US speaks in terms of producers and *their* participation in the certification process. Thus, we know that "FSC standards have been applied on more than 170 million acres of actively managed forest in more than 60 countries" (FSC-US 2006, 5), but we don't know how many dollars worth of certified products have been sold. FSC-Chile, on a much smaller scale, reproduces this effect; their website and newsletters (ICEFI 2007) document each new certificate awarded, but make no mention of units sold, profits or premiums earned. All of this begs the question—does indexing producer participation in certification tell us anything about the consumer side of the equation. Who's buying all this, and how much is produced and purchased?

Now, absence of evidence is not the same as evidence of absence, as I discuss below. Nonetheless, the gaps in FSC's literature point to a significant challenge in the effort to market certified wood products to consumers. I encountered this challenge in one of its oddest forms in Chile, and I'll ask the reader to put yourself in my position for a moment. You're in Santiago, sipping espressos at the tony Opera Café with a Chilean and a North American environmentalist. Recently, they abandoned their respective NGOs to form a team of "international sustainable marketing consultants." One consultant waxes indignant about the efforts that FSC-Chile has put into marketing the products they certify. "Did you know," he tells you, "that Chileans eat so-many liters of ice cream every year? All of it served on wooden sticks. Why aren't those sticks certified FSC wood? With a slogan about sustainability on each one! 'Taste the forest!' Every Chilean with the FSC in his mouth. Or tongue depressors! Or even better, what about swabs! Vaginal swabs! Imagine the symbolic value if we could capture that market." If you have trouble with the symbolic value of sustainable swabs, you are not alone.

But if the imagery is unfamiliar, the problem is not. The uptake of certified wood by consumers has necessarily followed a trajectory specific to its material qualities and production–consumption systems. Its commodity chains must be different, less consumer-driven, than comestibles like organic vegetables or Fair Trade coffee (Taylor 2005). In the end, even high-end sustainably grown wood requires a different consumer logic. There is nothing visceral about the consumption of wood products, no bodily hook on which to hang the necessary conflation of moral rectitude and Epicurean luxury (Lockie 2002). Those of us outside the building trades likely have little sense of the quality or provenance of wood products that we might purchase. Through the early part of the century, the FSC ran a series of print ads featuring celebrity testimonials, including musician Jennifer Lopez and actor Pierce Brosnan. Yet, it remains unclear if this advertising has generated market interest or created a widespread conscience of certification among consumers. It is not even clear what sort of audience the organization was attempting to reach.

All of which leads me to the issue of evidence, that is, evidence for a demand-side movement in support of certified wood. According to the two international sustainability consultants I met for coffee, according to every timber company forester I spoke with, according to the environmentalists, Chilean timber producers are not seeing a price premium for certified wood (also see CORMA 2007; FSC-Chile 2006). Regional markets have their own peculiarities, but Chile is not unique. In general there is no 'market' in the classical sense for certified wood; that is, nowhere do we find a significant number of consumers demonstrating their willingness to pay a significantly higher price for certified wood (Haener and Luckert 1998; Meidinger

2003). In short, it is *not* retail-end consumer demand and its price premium that is driving social change in forest certification. To find that motor we need to look further along the network.

THE ENVIRONMENTALIST

Since when do environmentalists sit down with loggers to devise the standards by which logging will be conducted? Some environmentalists undoubtedly feel that this kind of collaboration amounts to selling out, to handing over tracts of invaluable wilderness—to surrendering the fight. Consider a childhood anecdote related by Todd Paglia, executive director of Forest Ethics and an early contributor to the infrastructure of forest certification in Chile and elsewhere. Paglia describes a misadventure in which he and his young friends punished a supposed despoiler of the environment:

> And we knew nobody would do anything about it unless we did. So we came up with a plan . . . [we saw] what we were after—the guy's brand new Zodiac inflatable motor boat, it was the envy of the whole lake. We took the knives out of our mouths—we had swum across the cove with knives in our mouths—and we sunk his boat. Then we slunk back into the water, swam back across the cove, and . . . but before we left, we swept our tracks away so that they couldn't tell if the attack came by land or sea. We went back, got out of the water, and that was my first environmental act. It was an inauspicious start.

Paglia continues, "The strange thing is, looking back on that experience, there were parts of it that were in parallel to my early career, and uncomfortably so" (Paglia 2007). This charming tale of passage from youthful adventurism to mature engagement reads like a meta-narrative shared by many environmentalists throughout the movement. A dedicated scholar would find plenty of material for a "discourse and rhetoric" study on the tropes of maturity and self-control in environmental activism.[4] That same rhetoric traces the path of a process of institutionalization that has characterized the biography of world environmentalism since the 1980s—perhaps not coincidentally a similar timeframe as Paglia's personal microcosm.

To a certain extent, this dynamic falls under the broad heading of *institutionalization*. Political scientist Peter Haas coined the term "institutions for the Earth" (Haas et al. 1993) to describe the alphabet soup of governmental, para-governmental, and NGO bodies that began proliferating in the early 1980s. This proliferation is not just a scaling-up of popular concern about the environment to higher and higher levels of governance. Rather, it represents a qualitative transformation, from grassroots mobilization to the incorporation of environmentalism in the concerns of government itself. Incorporation of course is hardly transparent or total; what "the environment" and "environmentalism" can mean is transformed in the process.

The institutionalization of environmentalism follows an arc similar to that of international development, as documented by critical anthropologists like Ferguson (1990) or Escobar (1995). Ferguson called this pattern the 'anti-politics machine,' a technocratic structure by which the material needs of the less powerful are disarticulated from their access to power. Development produces "the poor" as targets for intervention, and constructs economic growth and government planning as the necessary and sufficient responses to inequality. By casting top-down intervention as an apolitical management technology rather than the political strategy of elites, development discourse inhibits the agency of the less powerful. Brosius (1999) considers how these dynamics play out in the realm of conservation and environmentalism. In particular he notes the potential for institutional legitimacy to constrain the public imagination from alternative forms of doing environmental politics:

> Such institutions, whatever else they may do, inscribe and naturalize certain discourses. While they create certain possibilities for ameliorating environmental degradation, they simultaneously preclude others. They privilege certain actors and marginalize others. Apparently designed to advance an environmental agenda, such institutions in fact often obstruct meaningful change through endless negotiation, legalistic evasion, compromising among 'stakeholders,' and the creation of unwieldy projects aimed at top-down environmental management. More importantly, however, they insinuate and naturalize a discourse that excludes moral or political imperatives in favor of indifferent bureaucratic and/or technoscientific forms of institutionally created and validated intervention. (Brosius 1999, 38)

The historical context of Chilean environmental activism adds an interesting wrinkle to this meta-narrative. On the one hand, the environmental organizations themselves offer a discourse of public transformation, in which they play a starring role in bringing the people of Chile to treasure their natural environment. This discourse in itself is noteworthy as a study in emergent environmental conscience to put up against that of Brosius's institutions or Agrawal's (2005) Indian foresters:

> Today if I tell the children that in 1968 whales were hunted and slaughtered in a business right off the coast of Concepción, they'd look at me like I was crazy. I wonder how the founders of CODEFF would look if someone had predicted that in just a few decades, the members of CODEFF could participate in our trips to see right whales in Argentina, and that the Argentines participate in visits to observe blue whales in the Gulf of Corcovado. In 1968, there was no CONAMA [Comisión

Nacional de Medio Ambiente], no CITES [Convention on International Trade in Endangered Species], no private parks, no forest certification, no wildlife rehabilitation centers. In 1968, CODEFF was born with much enthusiasm and optimism. It was time to bring to life an organization that would look out for our natural patrimony. After 36 years of fruitful labor, we can look back with pride. (CODEFF 1999)

Here we follow the convention of dating the birth of the Chilean environmental movement to 1968 when the Comité Nacional Pro Defensa de la Fauna y Flora (CODEFF) was founded. What does not appear in this capsule summary is the singular event that shaped Chilean history between 1968 and the present—the dictatorship of Augusto Pinochet. For Chile like much of the world, 1968 was a watershed year; it marked not just the foundation of CODEFF, but the radicalization of Chilean society both left and right. The first phases of institutional environmentalism in Chile emerged during a period of political tension, mass mobilization, and conflict that culminated when the democratically elected socialist government was toppled by the US-supported military coup on September 11, 1973 (see Loveman 2001).

Less eminent organizations than CODEFF came of age at the height of the dictatorship (Instituto de Ecología Political) or during the transition to democracy (Fundación Terram), and in each case demonstrate approaches and ideological commitments shaped by their relationship to the dictatorship. Three activist biographies illustrate this relationship: Luis Astorga, a forestry engineer and currently president of FSC-Chile, was *concientizado* [radicalized] as a forestry consultant in Central American in the early 1960s. He applied the grassroots ideologies and participatory forestry methodologies that he learned there on consultant projects throughout the global south until the mid-70s. When the military government began to pressure him, Astorga went into exile in Finland, and only returned to Chile after the transition to democracy.

Luis Otero, also a forestry engineer, was younger at the time of the military coup. A student activist, he spent time in secret military prisons on more than one occasion. We spent a good deal of time together, but one encounter stands out for me. We had scheduled an interview at his home in Valdivia some weeks earlier; the day before the interview, General Pinochet died in Santiago. In place of the interview, we sat late into the night watching television coverage of the dictator's life, the debates about his funeral, and the riotous celebrations in the streets of Santiago. Bernardo Reyes, cofounder of the Institute of Political Ecology and now an 'international sustainability consultant,' was also a student activist. He helped put together the Institute to provide a home for environmentalist voices more overtly critical of the dictatorship than older mainstream organizations like CODEFF.

When students of Chilean politics refer to the "legacy of the dictatorship" in environmental politics or elsewhere, they are typically referring to the set of institutional structures and public discourses that serve to inhibit mass participation in civil society (i.e. Carruthers 2001). For Paley, the relationship between grassroots mobilization and dictatorship appears across the spectrum of Chilean civil society. Her account of Chile's democratic transition (Paley 2001) documents citizen movements for improved community health care, neighborhood security, public transportation and housing "before and after" the dictatorship. In her estimation, all of these movements hit their peak of participation and effectiveness during the height of the dictatorship and its repression, and in the wake of the transition to democracy they largely dissolved. Chilean civil society in general after the return of constitutional government has ceded most of its energy and legitimacy to the mainstream political parties; journalists and politicians left and right often describe political activism outside of these contexts as nostalgic agitation or reckless endangerment of the fragile new democracy. This legacy of the dictatorship in Chilean civil society echoes the accounts of transformations in environmental activism in post-socialist Eastern Europe (Dawson 1996; Harper 2005). In referring to the personal lives of these environmental activists, I aim only to suggest that the legacy of the dictatorship in terms of memory and affect will also have its part in shaping the possibilities of environmental politics in the post-dictatorship period.

When the director of CODEFF describes the environmentalist assumptions of today's children in contrast to the state of environmental concern in 1968, he is invoking a discourse of green subject formation directly tied (even if by omission) to major geopolitical events. In the same fashion, the attention that environmental organizations pay to certification is a political statement on their part—they are not just doing regulation, they are themselves identifying with neoliberal discourse, and inserting themselves into a global imaginary that encompasses the forests they love, the nation they constitute, the movement they inhabit, and faceless Northern consumers they are attempting to enroll. Environmental NGOs are in the interesting position of acting as both goads to and proxies for these hypothetical sustainable consumers. Naturally, opinions within the NGO community are divided as to whether certification standards really measure the needs of ecosystems (Bennett 2001), and whether market incentives for sustainability can really outweigh the value of short-term profits (Carrere and Lohmann 1996). Still, as the worldwide acreage under FSC certification continues to expand so rapidly, it appears true that, as some of my Chilean informants told me, "whatever you think of the market, from here on it's pure FSC here in Chile."

Despite some obvious criticisms, the environmentalist embrace of market-based strategies does not mean a straightforward substitution of shopping for political activism. Just as the timber companies I discuss below pursue certification for motives beyond the potential price premium itself, environmental activists use certification as a market-based tool for political leverage. International organizations active in and about Chile, from ForestEthics to Greenpeace, have waged ongoing campaigns against big-box and catalog retailers like Home Depot and Victoria's Secret. Although these campaigns include targeted boycotts, letter-writing, and street theater as in the past, the expansion of forest certification has driven a subtle shift in their demands: That major corporate consumers of wood products "voluntarily" agree to purchase *only* certified wood (Carlton 2004; Lee et al. 2009). Retail companies are not the only target for this form of pressure: Members of FSC-Chile informed me excitedly about a recent triumph in which NGOs convinced the government of Spain to purchase only FSC-certified wood for government projects. Other institutions of environmental accountability provide another avenue for certification activism, as in the Leadership in Energy and Environmental Design (LEED) initiative of the US Green Building Council. To obtain the highest level of LEED's prestigious Green Building Certification, builders must use specifically FSC-certified wood products. With more than 20,000 buildings in 41 countries, that institutional element represents a significant hold on a certain market for FSC wood. Although LEED does provide residential certification, the majority of LEED construction is for offices, schools, government and institutional buildings (USGBC 2011). Purchasing decisions in these contexts are not individual ethical choices, but institutional ones, with all of the constraints and demands that institutions face and the pressures that activists can bring to bear on them.

As Chile's environmental organizations began redefining themselves in the post-dictatorship era, a diverse coalition was forming around a new way to promote sustainable forest management. The timeline in Appendix A shows the trajectory of FSC-International as an institution developing through the 1990s and the current decade. FSC didn't have a monopoly on the concept of forest certification, and indeed alternative programs proliferated throughout the late 1990s (Ozinga 2001; Cashore 2002). There are two aspects of the FSC's development that I find particularly interesting here—first, although certification was originally cast as a new way of imagining forest protection, it has been increasingly integrated with traditional conservation, as in WWF's Global Forest and Trade Network (White and Sarshar 2004). Second, FSC certification has been demonstrably more active in Northern temperate forests, rather than the Southern tropical forests for which it was initially conceived (Cashore et al. 2003; Colchester 2003; Elliot and Schlaepfer 2001).

FSC-Chile intersected this arc in the late 1990s. AIFBN as an organization was a member of FSC-International since 1995, at the behest of Luis Astorga who served as their representative to the international body as well. In 1998, Astorga organized a meeting with several Chilean organizations that brought then-director of FSC-International Timothy Synott to Chile. The primary actors behind this meeting were Astorga himself, already an active member of FSC-International; Hernan Verscheure with CODEFF; and Luis Otero, who had been working for several years as a SmartWood auditor on FSC-certified properties throughout Latin America. As Astorga put it, "*Esta malla social entonces presenta sus antecedentes a FSC internacional como iniciativa chilena del FSC, y logra que el FSC la reconozca, y la reconoce oficialmente.* [This social mesh then presented its background to FSC-International as the Chilean initiative of the FSC [ICEFI, Initiativa pro Certificación Forestal Independiente], and got the FSC to recognize it, to recognize it officially]."

ICEFI began working on a draft version of Chilean standards in 1999, a process that went on back-and-forth between Otero and the board as a whole for several years (see Chapter 5). A particularly critical moment in transition from ICEFI to FSC-Chile (that is, from an independent initiative to the national affiliate of the international organization) came in 2002, with the Forest Leadership Forum and Trade Fair in Atlanta. Several members of the Chilean organizations were in attendance as FSC associates, and repeated to me the influence of the trade fair on their conception of the FSC as a valuable path of entry into global markets for a new kind of timber product. By this time in late 2002 a number of small to mid-size timber companies in Chile were expressing interest in FSC certification, and seven had been certified under interim 'generic' standards for management or chain of custody (ICEFI 2002). The generic standards lasted for several years as the organization now called FSC-Chile hashed over the new and uniquely Chilean set of standards. The final version was sent to FSC-International to be ratified in 2006, and the approval arrived from the international body during the writing of this document. As I discuss further in Chapter 5, the national standards, and their associated institutional politics, articulate with the international spaces of the FSC in some strange ways.

TIMBER PRODUCERS

As I've noted already, forest certification developed as a system of promoting sustainable forestry during a period of widespread concern about logging and the destruction of rainforests. If environmentalists envisioned a tool for transforming the global timber trade, timber companies became involved with certification for a wider variety of strategic reasons. Since certification programs came online in the mid-1990s the state of the world's timber market has shifted substantially; certification itself is only one of several economic factors in that change. The new shape of the timber trade has significant implications for how certification will function and how companies will engage with it.

For most of the 20th century, the majority of wood consumed by the industrialized North was harvested from naturally forested areas in those industrialized countries (for example, the US, Canada, and Scandinavia countries). Because of the high costs of transportation, processing of wood products typically took place in or near forested landscapes; this is known as a regional resource-processing model (Perlin 1991). A transition began to take place in the second half of the 20th century, from this model to a more diverse and global pattern. Driven especially by Japanese timber companies and their political influence (Dauvergne 1997), the timber trade more and more consisted of low-cost harvest in tropical countries exporting raw materials for industrial processing in the global North. This transition and its staggering impact on tropical forests shaped the context for the rainforest crisis of the 1990s, and the global social movement that developed in response to it (Brosius 1999b). The global timber market has continued to evolve, however, and new dynamics are arising as the influence of Japanese companies declines. Market analysts (Bael and Sedjo 2006; Bowyer 2004; FAO 2007) identify four factors shaping the latest transitions in the global market: Trade liberalization, which makes it easier to shift industrial processing and production to developing countries with lower labor costs; tree plantations and the associated efficiencies of large-scale rationalized growth-and-harvest; technological innovations in processing wood products, composites, and engineered structural wood beams; and especially the role of what are often called 'emerging economies' (*Economist* 2006). This last includes Russia and China, both of which are rapidly industrializing and liberalizing their forest sectors. China is currently the world's fastest-growing economy, and has become a huge consumer of wood products to feed its export markets in manufactured goods, as well as for the domestic use of its growing consumer class (White et al. 2006, Kun et al. 2007).

Chile is a prime example of this latest transition in the global timber market. The variety of political and ecological factors discussed in Chapter 1 make Chile ideal for the new plantation-based model of production.[5] Since 2003, Chinese and Chilean businesspeople and politicians have been promoting trade relationships and direct investment in the forest sector (Pereira 2007). This relationship led to the official signing of a bilateral trade agreement between the two countries (China's first outside of Asia) at the 2005 Asia–Pacific Economic Cooperation summit in South Korea (Smith 2005). The agreement and its near-zero tariffs have already shown an impact: Of $3.7 billion dollars' worth of wood products exported, 14.5% went to China. That's still second to the US, which received 17.4% of Chile's exports, but the rate of change is significant: The US market declined by nearly 25%, while the Chinese market rose 122% (INFOR 2007).

The implications of this market dynamic for forest certification are far from clear. To date, most certified forests are found in precisely those industrialized Northern countries that are contributing significantly less to global forest production (Bowyer et al. 2004). Meanwhile, what demand exists for certified wood products comes primarily from the US and western Europe (Cashore 2002), which make up a steadily smaller and smaller portion of the market relative to China and other emerging economies. Until and unless the Chinese market develops an interest in certified wood products, this would seem to make forest certification a relatively marginal project in Chile. The export market is dominated by two companies, CMPC and Arauco, neither of which have investigated FSC certification at all; Masisa, the only certifying company to compete at that level, still exports a fraction of what the two giants can produce (INFOR 2006). That being the case, what can certification do in Chile? Why should a company get involved? These questions were very much on the mind of the timber company officials I spoke with (also see CORMA 2007), and below I explore some of the issues that arise in regards to timber companies choosing to pursue certification.

During the first phase of my fieldwork in 2004, I conducted interviews with a number of prominent Chilean environmentalists. They reprimanded me on more than one occasion for referring to their boycotting and letter-writing and so forth as a campaign against the timber companies. "No, no," I was told, "this is a joint campaign, *together* with the timber companies."

The major innovation of forest certification, from the environmental NGO perspective, is the enrollment of timber producers as allies in forest conservation (Elliot and Schlaepfer 2001). For their part, timber producers seem to see forest certification as an opportunity to codify publicly their commitment to responsible forestry, without committing to binding expectations that they cannot afford to meet. Producers and retailers alike frequently claim that meeting FSC standards demands little or no change in their current practices (Carlton 2004; Hagerty 1999; also see Newsom et al. 2005). Nonetheless, the most common outcome when FSC certification arrives in a given timber market has been the proliferation of corporate-sponsored alternative certification programs—the Pan-European Forest Certification scheme, the Sustainable Forestry Initiative of the American Forest & Paper Association, and the Canadian Standards Association's Sustainable Forest Management Standard, and so on (Cashore et al. 2004). These initiatives tend to be much more producer-friendly than the Forest Stewardship Council, including process-based rather than outcome-based standards and first-party (i.e. self-assessment) rather than third-party certification. Related research on ethical and ecological standards in the cut-flower industry, for example (Hughes 2000, 2001), suggests that for business interests, participating in certification is essentially a form of risk-management, a containment strategy to limit their vulnerability to the claims of hostile NGOs. The lowering of the bar implied by self-assessment in corporate-sponsored certification initiatives seems to confirm this analysis. What does it mean, then, for timber producers to involve themselves with a body as relatively radical as the Forest Stewardship Council? Understanding these decisions naturally requires a certain backgrounding in the history of Chilean forestry.

The cutting of trees on a grand scale began in Chile, as in much of Latin America, with clearing for agriculture in the early colonial era.[6] Early on, however, Chile developed a large internal wood market, based on the constant need for strong timbers to support the expanding copper mines in the northern part of the country. As the independent nation's economy grew through the 19th century, Chile's forest sector entered the export market, sending sawn logs to other Latin American countries and even to the US for railroad ties and other industrial raw materials. A significant aspect of Chilean agrarian economy has tended to shape forestry trends since that time; land in Chile is overwhelmingly privately owned. Chile has no system of communal property ownership equivalent to Mexico's *ejidos*; there is no large body of government land holding equivalent to the Bureau of Land Management in the US; and the government does not sell concessions for timber harvest on public land as is so often done elsewhere in the global South. Aside from small units of recognized indigenous territory and the relatively recent system of protected areas (SNASPE, Sistema Nacional de Areas Silvestres Protegidas [National System of Protected Wild Areas]), forests in Chile belong to individuals and corporations.

Although it predates Chile's recognition as a neoliberal 'economic miracle' (see Loveman 2001), this system of private ownership has made aspects of the neoliberal forestry model especially functional in Chile (Clapp 1998). A second, ecological factor has played a part in making Chile's forest sector such a growth industry—it's been described as the "unnatural history of the Monterrey pine" (Clapp 1995). As I noted in Chapter 1, Chilean forestry depends almost entirely on plantation production; about 75% of the timber plantations are owned by 'large corporations,' nearly 50% by the two largest (INFOR 2005). What's crucial here is that the whole plantation economy would never have been possible if not for a fancy bit of global botanical trickery. Chile's plantations were built almost entirely from a single species, *pinus radiata* or the Monterrey pine. In its native California, the Monterrey pine is confined to a couple of very small mountain ridges, but scientific forestry of the early 1900s found much more efficient uses for it. In Chile's Mediterranean and cool temperate coastal regions, the same tree can grow to harvestable maturity within seven years, allowing for some of the fastest rotations in forestry worldwide. It's immune to the pest species that make native timber harvests unprofitable in Chile, and the pine bark beetle that limits its distribution in California and other parts of South America is walled out of Chile by the Atacama Desert to the north and the Andes to the east.[7]

The political ecology of private ownership and exotic species plantations plays a key role in how timber producers encounter a private system for regulating forestry. The questions raised by certification's virtual consumers still persist, though. In the absence of price premiums, why do timber producers pursue certification? Three distinct motivations appeared to drive certifiers, each with its particular way of incorporating and reinforcing the global discourse of certification.

First, most timber companies involved with FSC-Chile were concerned with protecting their corporate reputation. In the context of the ongoing territorial conflicts with Mapuche communities and the rising national concern about native forests, not to mention recent global interest in deforestation as a factor in global warming, timber companies were getting tired of being the bad guy. In interviews, I had several forestry engineers tell me that in fact, certifying with FSC would require no major changes in their company's already sustainable practices. I have no way to judge the accuracy of the argument at the moment,[8] but I see them in the context of other claims the timber industry has made to be good stewards of the environment and the 'real environmentalists.' Even when no direct price premium hangs on a relatively green corporate image, few companies can afford to be seen as environmental ogres, and many find the costs associated with certification worthwhile simply to avoid the bad press (see for example Carlton 2004). Certification with a widely recognized program like the FSC serves as an effective way to forestall environmentalist arguments about the quality of a company's practices. There seemed to be an edge of resentment in the language some foresters used to phrase this argument: "Because we the loggers have been demonized, and FSC has the approval of the environmentalists, now we can say that we are also environmentalists!"

A second important factor at work in timber company participation was the possibility of new markets, especially in Europe. Especially significant here is the development of new product lines to accompany the newly certified wood products; several producers were beginning to experiment with wood flooring, window frames, pre-cut trim, and other value-added products. This takes place in a market that has been overwhelmingly dominated by pulpwood, and represents a significant amount of risk. Corporations undeniably make strategic decisions like this in a larger context than simple analyses of the price premiums or cost-benefit ratios, as in the adoption of 'best practices' or 'corporate fads' (DiMaggio and Powell 1983) or the current interest in 'corporate social responsibility' (Sen and Battacharya 2001; Mohr et al. 2001). In Carrier's terms,

> If conformity to the current conception of good business practice does help to make a business succeed, there is reason to suspect that it does not do so simply because of the overt reason, that conformity to good practice makes a firm objectively more efficient. Rather, it may facilitate success because conformity to the standards of good business practice is a sign that the firm is a good business, a sign that makes it more likely that other firms will be willing to do business with it. To put this more conservatively, the failure to conform to good business practice is likely to be a sign that the firm is unreliable and unpredictable, one to be avoided and hence one that is likely to fail. (Carrier 1997, 13)

There is no particular evidence that these products would gain a significant premium in the foreign markets for having been certified, but several producers were gambling that certification might provide the edge to get them in (Bluth Solari 2002). The head of the certification project with Masisa, the largest company involved with FSC-Chile, told me: "Yes, it costs us to certify, with the dues and . . . There's a lot of paperwork to it, it costs a lot of time. But every strategy has a cost. The question is, can we afford not to, not to try? If the timber market keeps changing and we want to do it later, then we're—now can we afford it?"

Which leads to the third motivation for certification, a particularly complex articulation with the politics of global markets and environmental activism. Several foresters expressed concern that at some point in the future, markets that they currently rely on would be begin to demand *only* certified wood. In other words, they faced the prospect that FSC certification would make the leap from a not-actually-lucrative niche market to being the minimum standard for larger, mainstream markets. This is a concern that forest economists have been discussing since early on in the FSC's history; they refer to it as the "unrewarded minimum standard." Economists have a number of macroeconomic concerns about what such a standard would do to the forest industry in terms of promoting concentration and mergers (Haener and Luckert 1998); the producers I spoke with were simply worried about having to take on greater costs for the same return. They felt that adopting the process early and adapting their forestry practices gradually and with the support of the FSC and its member organization would make the transition easier. As noted above, environmental activists are working toward much the same goal; as these two parties in the FSC domain tug in opposite directions, they are gradually generating a new set of approaches to the global timber market.

An element of the Chilean situation that complicates this further is the vast gap between large and small timber producers. Since 2005, CODEFF has been sponsoring the development of a collective certification project targeting small proprietors in the remote rural community of Curarrehue in a mountainous part of Chile's IX region. The organization, called Mawidakom after a term in the Mapuche language, *Mapudungun*, that means "living with the forest," is made up of Mapuche and other Chilean owners of properties generally under 30 hectares. Most are engaged in a general mix of small livestock, subsistence and local market vegetable crops, and handicrafts for the tourist market in nearby Pucon. The process of bringing together enough of these participants to create the conditions for group certification has been slow and halting. It has involved a great deal of back-and-forth action between CODEFF, its funders in charitable foundations, the forestry engineer they contracted to be their extension agent, and different factions within the community organization itself. A consistent problem, though, has been that FSC's requirements for certification, like Chile's timber laws in general, were written with the impacts and resources of large- or medium- scale industrial producers in mind. One response to this problem has been the creation of a sort of manual for small producers on which I collaborated, translating the legalistic language and big numbers of Chilean forest law into terms more relevant to small rural producers. Even with small supports like this, it is clear that the capacities and interests of small producers, especially Mapuche, differ dramatically from those of even relatively small commercial producers. A conversation with one Mawidakom member threw this into relief:

> I give thanks to God every day that the forest, for me, it's my savings account. When I don't have work, then I head to the forest and cut something for, for . . . to save on the basic needs of the house. And the kids in the neighborhood too. When I talk to them, the first thing we talk about is the forest, now I'm even buying a book to study natural medicines, because in these forests there are so many remedies. I hope that others do the same, because the forest is the nicest thing we have.

To the extent that it takes its mission to promote sustainability throughout the Chilean forest sector serious, FSC's challenge is to create an institution that can speak to the situations of large producers and tiny ones alike. As this same Mawidakom member told me when I asked how she became interested in certification in the first place:

> Well, because they told me lots of great stories. Because it said if we certify the forest, we'll be able to sell, to sell where we want. That sounded like something good, I said, "Good, good, a tree that I sell now, they pay me maybe 20,000 [pesos, about $40], better if they're going to pay 50,000 or even 30,000, it's a little more. But no, there's still no document like that, nothing, and we've got no claims to make.

Like larger producers, small producers like this one need to find reasons to make certification effective for them. But the politics of scale that I discuss in more detail in Chapter 5 are material and present for the small producers in a way that timber companies simply don't experience. In the process of managing these challenges, FSC-Chile finds itself spread thinner and thinner.

THE MAPUCHE

When I set out to interview participants in the formation of FSC-Chile as an institution, it became clear that I would need to speak not just with members of the board and so forth, but also with actors *not* participating, for one reason or another. Primary in that group of non-participants were Mapuche activists, who fell outside the ambit of the FSC both by neglect and by choice. I had an interview with a noted and militant Mapuche activist who was particularly noteworthy in that respect. As part of my usual 'informed consent' routine, I explained at the outset of the interview that he could end the interview at any time, or could choose not to answer any question for any reason he saw fit. Taking me at my word, he politely declined to answer most of my questions, even the most basic:

Q: Can you tell me where you were educated?
A: We can't answer that.
Q: Your organization's website says that you have your own plans for sustainable development independent of the government. Can you tell me a little about that?
A: We can't answer that.

And so on. But when I brought up the subject of relations between Mapuche activist organizations and Chilean environmental organizations, he practically exploded:

> Let me tell you, the environmentalists, we call them computer warriors, they sit behind the computers and never do anything concrete. We're involved in an environmental struggle, a territorial struggle, and there are lots of environmental institutions that all they do is justify projects from behind their computers. They don't have a clear position on free trade, they don't have a clear position on natural resources . . . Definitely, the environmentalists are fine idealists but they're not concrete activists. The Mapuche struggle is an active struggle in the field, on the land, and to me that's what's valid. That other is *acceptable*, but it's not right.

None of the other Mapuche activists I spoke with expressed quite as much rage towards environmentalists—but many chuckled with recognition at the epithet "computer warriors," and many expressed a lack of confidence that environmentalists were reliable allies for the Mapuche movement (also see Tockman 2004; Henne and Gabrielson forthcoming). It's a theme that came up repeatedly when I spoke to Mapuche about their feelings on the FSC; that even when they made common cause with environmentalists, the relationship was marked by the same dynamics that defined the place of the Mapuche in Chilean society more generally.

What that place is, of course, is not always clear, and it has been shifting with the political climate of Chile for centuries. A recent controversy illustrates the scope of the problem. In 1992, the government census measured ethnicity in the Chilean population for the first time (INE 1992), and the country was astonished to learn that the Mapuche population was estimated to be nearly one million, or almost 10% of the national total. Half that number lived in the south, in traditionally Mapuche territory, while the other half lived in the greater metropolitan region around Santiago. The numbers came at a turbulent time; the post-dictatorship "return to democracy" (Paley 2001) overlapped with the five-hundredth anniversary of Columbus's arrival in the Americas and the worldwide attention to the indigenous Americas that came with it (Bengoa 2000b). The 1992 census stirred up a nationwide discussion about the place and status of indigenous populations in Chile (Aylwin et al. 2004) and contributed to new policies in terms of targeting educational funding, land reform, and development projects (Castro Lucic 2005).

Given Chile's history of colonialism and dictatorship, the post-1992 stance represented something of a sea-change. Throughout the Pinochet dictatorship, public discourse suggested that Mapuche culture was a relic, that in a modern society "we're all Chileans here" (Pinto 2003). One Chilean sociologist has a best-selling monograph on the subject, in which he denies the authenticity of most Chileans claiming Mapuche identity; in fact, he essentially argues that the Mapuche strictly speaking no longer exist, having effectively merged into the greater pool of Chilean citizenry nearly a century ago (Saavedra 2000). An extreme version of this discourse persists in Chile and elsewhere. I once gave a presentation on this research to a forest agronomy class in the US, taught by an economist who had worked as a consultant in Chile while the military dictatorship was actively promoting and subsidizing new plantations. At the end of my talk about the plantation economy's impact on indigenous communities, he raised his hand and said, "But, there *aren't* any indigenous people in Chile." The 1992 census made this idea untenable.

In 2002, however, the government agency conducted a new census (INE 2002). The new numbers were radically different; rather than a million Mapuche, the census showed barely 600,000. A number of factors make the two figures difficult to compare: The age-blocks in the two studies differed, as well as the options for ethnic identity (the later census offered eight different indigenous ethnicities rather than one). Most tellingly, the 1992 census asked "Do you identify as indigenous?" while the 2002 census asked "Do you belong to one of the following [indigenous] communities?" Many speculate that, as many Mapuche are residents of urban centers, the question of 'belonging' to a 'community' may have dramatically skewed the results. Given the political capital that the Mapuche movement had gained from the 1992 census, there was an outcry in response to the new numbers, including accusations of "statistical genocide" (Bañados and Frias 2003). To date, neither the government nor the civil society of Chile can agree on the number of Mapuche or how even to find out. In the end, the debate reflects basic disagreement not only over the number of Mapuche, but their existence, and their collective or cultural rights as a people.

The census debate appeared as one more in a centuries-long series of political conflicts. Even before the Spanish arrived, the ancestors of the Mapuche were engaged in a battle of resistance against the southward expansion of the Inca Empire. Nomadic and decentralized, the Mapuche proved impossible for the faltering empire to subdue.[9] When Pedro de Valdivia first entered the territory now known as Chile in 1541, he established the city of Santiago as a military outpost of the colony in Peru. Santiago, as well as the other settlements Valdivia's army established in their push to the south, was razed to the ground on more than one occasion by the Mapuche resistance. The Treaty of Quillin, signed in 1640, established nation-state relations between Spain and the Mapuche, with a border defined at the Bio-Bio River. The long period following the Treaty of Quillin resembles the westward expansion of the US frontier in many ways. While recognizing the sovereignty and territory of the Mapuche in a sequence of treaties and agreements, the Spanish continued to push the border southward, capturing slaves and destroying settlements (Bengoa 2000).

The endless war was an enormous strain on the resources of the Spanish Crown, and many historians argue that the resistance was a key factor in enabling Chile's successful bid for independence (Mariman et al. 2006). With Chilean independence established in 1818, the military turned its attention completely towards eliminating the Mapuche resistance, a campaign euphemistically referred to in Chilean history texts as the "Pacification of Auracania," and which the Mapuche call the "War to the Death in the South" or "The Last Massacre" (Bengoa 2000). The last large-scale Mapuche uprising ended in 1881, when the Treaty of Nielol was signed in Temuco. Mapuche historians are at pains to point out that the terms of the treaty were not those of a surrender, and that the Mapuche resistance was never defeated as such (Marimán et al. 2006).

Nonetheless, the terms of the peace finally broke up Mapuche sovereignty south of the Bio-Bio, and most of the populations were confined to *reducciónes*, very limited lands defined by the state as communally owned indigenous territories. As Louis Faron documented in one of the few classic ethnographies of Mapuche *reducción* society (1986[1940]), the Mapuche communities were confined to marginal lands, deprived of much of their livestock as well as the freedom to graze in highlands and trade across the Andes. Very rapidly, they went from a relatively affluent society of highly mobile pastoralists and livestock traders to a destitute rural peasantry.[10]

Territory and land tenure under the *reducción* system was a messy patchwork from the beginning. The treaties signed by the Chilean state in various times and places were not necessarily honored in the construction of particular communities, nor were the remaining *titulos de merced* (mercy titles) that had been granted by the Spanish Crown. In many cases Mapuche communities or individuals petitioned to Chilean courts to secure lands they had been granted by earlier agreements; many times this intervention failed (Pinto 2003; Frias 2003). Mapuche communities participated in large numbers in the collective struggle for land reform in the late 1960s and early 1970s, often in collaboration with *campesino* cooperatives with little acknowledgement of their identity as indigenous (Mallon 2005). As Frias (2003) and Mallon (2005) document, the Pinochet dictatorship targeted Mapuche communities during their "rollback" of the agrarian reforms, and most of the territories claimed (or reclaimed) by communities were nationalized and re-privatized (Klubock 2004). The great beneficiary of this process was the timber industry, and the suddenly available arable land played a key role in the Chilean timber industry's rapid growth through the 1980s (Clapp 1995a).

In the post-dictatorship era, perhaps bolstered by the 1992 census and the rise of hemispheric indigenous activism (Bengoa 2000b), Mapuche communities and activist organizations again began to make concerted efforts to reclaim territory. Timber companies with large holdings in the south became the primary targets for this movement, which ranged from various forms of litigation to boycotts to land seizures (Rohter 2004). The authorities responded to the increasingly militant Mapuche movement with a level of violence that startled Chilean civil society (Aylwin 1998), including violent police crackdowns that have led to the deaths of at least two activists and the injury or imprisonment of many more (Witte 2008). Many of those imprisoned for participating in protests or land seizures have been prosecuted under anti-terrorist laws established by the Pinochet dictatorship. According to these laws, crimes constitute terrorism if they are intended to "spread fear in the population or part of it," whether violence took place or not, and such crimes bear much harsher sentences than ordinary crimes.[11] In Chapter 5 I discuss in more detail how these land conflicts between Mapuche communities and their Chilean neighbors intersect with the agenda of the Forest Stewardship Council.

The conflict over forestry and land claims in Mapuche territory is accompanied by a dramatic increase in Mapuche cultural activism, a revitalization movement that has been likened to a 'great awakening' (Aylwin et al. 2004). A significant aspect of the growth in Mapuche cultural awareness since the mid-1990s has been the successful regional program for bilingual education in Spanish and Mapudungun for indigenous students (Marimán 1997), as well as the proliferation of bicultural health care centers where certified practitioners of Mapuche traditional medicine collaborate with medical professionals (Richards 2006). In the face of international interest in Mapuche culture and rights, as well as the controversy over the 2002 census, Chile's national discourse on indigeneity remains troubled (Richards 2007).

On the face of it, FSC certification could work to the advantage of Mapuche activists in their attempts to enroll support for the territorial claims:

PRINCIPLE 3: INDIGENOUS PEOPLE'S RIGHTS

The legal and customary rights of indigenous peoples to own, use and manage their lands, territories, and resources shall be recognized and respected.

Criteria:

3.1 Indigenous peoples shall control forest management on their lands and territories unless they delegate control with free and informed consent to other agencies.

> 3.2 Forest management shall not threaten or diminish, either directly or indirectly, the resources or tenure rights of indigenous peoples.
> 3.3 Sites of special cultural, ecological, economic or religious significance to indigenous peoples shall be clearly identified in cooperation with such peoples, and recognized and protected by forest managers.
> 3.4 Indigenous peoples shall be compensated for the application of their traditional knowledge regarding the use of forest species or management systems in forest operations. This compensation shall be formally agreed upon with their free and informed consent before forest operations commence.

This is the statement of FSC's Principle 3 at the level of FSC-International. Until the local or subregional body defines specific criteria, indicators, and verifiers, these International Principles and Criteria can be applied as 'generic standards.' It is in the developing the local standard that FSC-Chile gains meaning, hence the expectation that standards will be negotiated by a representative roundtable of stakeholders. This is why, although the generic statements are probably a stronger statement of indigenous rights than any other language to be found in Chilean environmentalism, the end result is less than satisfactory to Mapuche activists.

The issue of representation in the negotiating process follows a familiar pattern. Rhetorically, the framing of Mapuche territory in relation to forestry has a great deal to do with citizenship. As I've noted already, timber companies across the board represent themselves as the great hope for Chilean development, stewards of a nearly inexhaustible resource for lifting a fragile democracy up from poverty and dictatorship. Questioning the thesis for whatever purpose gets one labeled a communist and a traitor.[12] Interestingly, even as they dissent from it, environmentalists reinscribe this dominant discourse of forests and citizenship. The biodiverse rainforests, they argue, are the patrimony of *all Chileans*. This is a discursive category, *all Chileans*, that explicitly writes out cultural or economic differences within and among Chileans—such as Mapuche ethnic identity. This can be true even when environmental actors are clearly sympathetic to the Mapuche struggle for recognition: As one environmental activist and FSC member told me:

> We not only understand the Mapuche suggestions, but we're very concerned about what's happening to them. But we can't make an association with the Mapuche directly because the environmental organizations that are involved in this campaign are truly doing it for the entire country, while the Mapuche class is located in the IX, X Regions. . . . I suspect that to support the Mapuche demands would really take ourselves out of the game.[13]

Making Wood and Making Persons 53

As I combed through the minutes of the hundreds of meetings that FSC-Chile conducted, I found a consistent ambivalence about the issue of Mapuche territory. Generally everyone, timber companies included, agreed that it was important to put down in black-and-white that no property taken from indigenous communities should be certified, and that certification should be a sign of clear title and good territorial relations. But very little is ever said about defining and enforcing that standard. All parties to the discussion, timber companies and environmentalists alike, appeared to be comfortable with the idea that the limits of Mapuche territory lay where the state had drawn them. There simply isn't any contention or alternative viewpoint expressed on record anywhere.

I brought this startling point up with one of my key contacts, Pablo Huaiquilao, a young Mapuche activist and forestry engineer who happened to be FSC-Chile's one and only Mapuche representative. Pablo's experience of the process by which the Chilean standards were negotiated stood at stark odds with the accounts of the non-Mapuche members, even those with whom he had worked closely. He told me that he had joined the board at FSC after the discussions about Principle 3 and territory had been completed and the drafts ratified:

> A: The other important theme, then, about timber conflicts, and the question of territory. I still don't have a good picture of about how that drives certification. Principle 3 says that a certified project must respect indigenous territory, but how was that defined?
> P: That's a controversial subject. Because there are several timber companies that have had problems with communities. And they hide in the papers, the papers are a, are an accidental tool that don't have validity as customary law, of indigenous territory. So they are, it's a slanted vision of justice, to say it another way. I mean, if this one has rights, because there's a signed paper, and this one doesn't have rights because there's no signed paper. Today certification lives by that, by those papers. And that's a really troubling subject.
> A: So during negotiating the standards . . .
> P: So, the problem is that I joined after the discussion on this point. So this point was sanctioned when I arrived, unfortunately. And unfortunately, the cases that have been given, there isn't much to do, because, the companies can't, they didn't fix the certification, because they have legal papers in their favor. The indigenous law, the customary law of possession, of the Mapuches, is much older that the rights written in these papers. So there you have something that has to be revised, defined. But I think that in this process of the certification the standards are constantly being revised, and I think that there we're going to open the discussion again, so it can be much more defined and reflect in some manner the concerns within the Mapuche communities and organizations.

54 *Making Wood and Making Persons*

I found it startling that even in the midst of an interview in which he spared the FSC no criticism, on a subject of huge importance for communities he's spent years supporting, Huaiquilao still chose to finish on a positive note. This suggests, to me, that despite false starts the relationship between Mapuche territorial rights and environmental regulation will continue to evolve.

I trace these subject positions not in the interest of positively identifying the 'real' dominant actors in a self-existing framework, nor am I trying to locate potential stakeholders for forest certification. Instead, I am interested in the way the logic of market-based social change initiatives like the FSC calls particular social collectives into being in particular configurations. How those collectives interact, and how their collective representations of/interventions in the world's forests might happen to resonate or collide—these are the processes by which individual acts of interpellation become technologies of value, become social configurations with the power to influence environmental regulation and the ways that forests are managed. In the chapters to come, I take on the knowledge practices that make nature complicit in these person-making projects.

NOTES

1. From Neil Smith's *Uneven Development* (1984) to Anna Tsing's *Friction* (2005) or the recent multispecies turn (Helmreich and Kirksey 2010), and quite a lot in between (i.e. Braun 2002; Castree 1995; Cronon 1995; Escobar 1995; Haraway 1991; Hayden 2003; Raffles 2002a, etc.).
2. While I can't yet say what implications this has for the functioning of FSC-International, I should note that the Canadian regional FSC affiliate has already instituted a fourth body, the Indigenous chamber, thereby troubling some of the reifications while suggesting some interesting new ones (Tollefson 2004).
3. Although advocates certainly make the effort to do so—see for example the Organic Consumers Association at www.organicconsumers.org.
4. In the meantime, see Satterfield's (2002) *Anatomy of a Conflict*, which does consider the oddly contradictory roles of emotion and self-control in forest politics.
5. See Clapp (1995b) for a discussion of just how complicated and ugly such a process can be.
6. Although see Otero (2006) for a consideration of the depth and breadth of forest clearing by Mapuche and other people prior to colonization.
7. Many foresters told me that with the warmer winters of recent years, the beetle seems to be edging further across the Andes. Some speculate that the whole Chilean plantation sector may end up a casualty of global warming. All this to say that humans are not the only ones in the forest sector changing with the times at a global scale.
8. Although see Newsome et al. (2005), who document the changes in practice required of US foresters on adopting FSC standards.
9. The name for the Mapuche used by Spanish colonists for centuries was *Auracanians*, reportedly derived from the Quechua [Inca] word for "rebel" (Loveman 2001).

10. Also see Pinto (2003) on the affluence of Mapuche traders well into the late 1800s.
11. Up to and including indefinite incarceration. Standards of evidence are different in these trials as well, in which *testigos sin rostros* [faceless witnesses] may testify (Stavenhagen 2003).
12. Even one of Chile's most internationally famous personalities, novelist Isabel Allende, was not immune to this; her support for a timber campaign in 2002 sparked enormous outrage (Gonzalez 2002).
13. For a book-length treatment of this comment, see Sierra (1992).

3   Putting Knowledge to Work

BUILDING CAPACITY

The first thing that I notice about the offices of Forestal Rio Largo was the flooring—wide planks of the same luminous pink-red wood that I've been burning in the woodstove lately. They look lovely in the rambling old colonial house that accommodates the timber company's small suite of offices, and they give me a decent question to break the ice with the loggers filing in with me and shaking the rain from their sweaters. "What is this wood?" It's *ulmo*, and they look at me strangely—what's a gringo with shaky Spanish doing here? But the entire staff of the small company is gathering, and soon we're all dragging chairs into the hallway that passes for a meeting room. When I came by yesterday to interview Rio Largo's forester, Luisa Catalano, she invited me to come back out for tonight's training.[1] As the liaison for FSC at Rio Largo, Luisa's job is to ensure that the company is conducting its forestry activities according to FSC's standards, as outlined in their accreditation papers and *plan de manejo*. Tonight she is offering a training for all of the "field staff" (*empleos de terreno*) on FSC standards and certification. I have come to get a sense of what FSC certification means to the forest workers who actually plant the seedlings and cut the trees, and especially to see how an interlocutor like Luisa frames the issue to communicate it to them.

Rio Largo is in Lanco, a tiny *comuna* in the very north of the *provincia* of Valdivia more than an hour's bus ride from Valdivia itself. The old house is not in the best condition, and the building combines with the informality of the staff to convey the sense of a small, familial company operating below the margin of enormous corporate timber interests. They fill a small market niche, producing the wood for manufactured windows and pre-fab flooring materials for domestic and international markets—samples of their products are propped in corners around the offices.

When I arrive, Luisa is setting up a projection screen and laptop computer at the front of the presentation space, while workers trickle in and chat. There are some jokes about the operation running on "Chilean time," an incongruity as we're only about 10 minutes behind and Chileans in general have a near-German reputation for promptness. In attendance, eventually, are 17 people—12 men dressed in flannel and heavy sweaters, five women in the Chilean version of 'business casual.' They cover a range of ages, and among the older men are several whose dark skin marks them as clearly of Mapuche ancestry.

Luisa calls cheerfully for quiet and begins her PowerPoint presentation. When the FSC banner appears in English, she reads it with an exaggerated stumble: "Forest fuf-uf uf-uf-uf," as if to say "It's all Greek to me"—an odd gesture as I know she speaks English, but it's good for a laugh with this audience. She describes certification as a guarantee, like a diploma, which says that responsible management really did take place. This is to *tranquilizar* [reassure] consumers and the producers of products using their wood. Her vocabulary here is basic, her tone somewhere between casual and didactic, as though she were speaking to a class of junior-high students. She uses a lot of rhetorical questions.

"*¿Y que es el FSC? Son gran pensadores, ecologos, empresarios, antropologos, hippies.*" [What is FSC? They are big thinkers, environmentalists, businessmen, anthropologists, hippies.] The big thinkers thought and thought, and the outcome of all that big thought was 10 principles for *manejo sustentable* [sustainable management]. The 10 principles (which she has not yet defined) are a big vision, and it takes the criteria and indicators to make them concrete. Concrete is a word that comes up a lot in this presentation. Luisa goes on to describe the process of FSC accrediting auditors, naming the individual auditors who recently evaluated Rio Largo and nodding to employees who apparently would have had interactions with them. At this point she stops for a question-and-answer session, calling on employees by name. "*¿Que es el manejo sustentable?*" [What is sustainable management?] Her audience, while friendly, is clearly reluctant to speak up. Some answers included 'doing a good job, being careful—it's good forestry,' and 'it's a process.'

The next question, "*¿Que es la cadena de custodia?*" [What is the chain of custody] is directed to me, as we had discussed the topic yesterday, but I demur. Other answers include *un rastro* [trail or trace] and *trazabilidad* [traceability], a bit of bureaucratic vocabulary that makes Luisa raise her eyebrows appreciatively. She replies with a depiction of a consumer in the 'Home Center' who sees a sack of firewood, " *y puede ver, mira, este leña viene de este empresa y este unidad en este año, por eso puedo sentir seguro que compro un producto bien manejado*" [and they can see, well I see that this wood came from this company and this lot in this year, so I can feel safe that I'm buying a well-managed product]. To this, one of the female office staff says, "Oh, it's information."

Luisa then moves into a discussion of the principles "*en practica*;" making the Principles practical she defines as the task of the criteria and indicators. At this point she goes through the Principles one by one with PowerPoint slides illustrating what kind of practices indicate putting the

58 *Putting Knowledge to Work*

Principle to work. That is, Principle 4 regarding the security of workers is illustrated by pictures of safety equipment, pages from manuals for training people in its use, graphics depicting proper felling techniques. In this fashion Luisa moves piecemeal through the Principles: Aspects of Principle 5 associated with "enhancing the value of forest services" are demonstrated with images of workers planting native species; "controlled use of chemicals" in Principle 6 apparently means using them only when necessary and with full safety gear, and so on. I note that Luisa shares considerably less detail on protecting wildlife or species diversity, as required by Principle 6 and 9. Also significant is the fact that in the presentation, social principles like indigenous territory (Principle 3) and the rights of neighboring communities (Principle 4) are addressed in terms of *"grupos de interes"* [interest groups], illustrated by a photo of a parent/student group from a nearby high school touring a harvest site. *"Inversion infrastructural"* [infrastructural investment] from Principle 6 appears in the slideshow in the form of roads and bridges—Luisa asks the audience why this is important. The only real response is about worker safety, but she adds that without good roads, managers are more likely to neglect remote properties. Hence, a good road means good management. This approach to infrastructure catches my attention, as everything I have heard about local infrastructure up to now has referred to the risks and impacts of enormous, speeding logging trucks on roads that neighboring communities also use. Luisa's presentation of infrastructure and "interest groups" raises some issues about the relationship between a company and its neighbors, and indeed about how FSC standards work to produce "the neighboring community" in the first place.

The issue of clearcutting, which is addressed in Principles 5 and 6, creates a little confusion. Luisa comments offhandedly that clearcutting is forbidden under FSC standards, and this provokes a couple of snorts from the audience. *"¡Pucha! Siempre lo hacemos, po!"* [Well, shit! We do it all the time!] There is some discussion to the effect that the anthropological eavesdropper, having heard this dirty secret, must now be killed. Luisa hastens to explain that the definition of clearcut (*tala raja*) is a matter of scale: *Big* clearcuts are forbidden, little patchworks of clear-cut stands are okay, as are the variations of patchwork harvesting that Rio Largo employs—the *corta protección, raleo, corta faja*, and so on. At this point Luisa turns toward what seems like familiar territory, a general description of their management approach. Rio Largo prides itself on an integrated use of replanted native forest and exotic plantation in a mosaic that can provide wood products without threatening biodiversity. This patchwork approach gives them a lot of flexibility in terms of their crop rotations and harvest schedule, because *"¡Roble es precioso, lingue es precioso, pero roble tiene un rotacion de 40 años!"* [beech is lovely, lingue is lovely, but beech has a rotation of 40 years!] Luisa characterizes this pattern of forestry as a simple set of responsible decisions based on sound economic principles. So why certify it? *"Porque nosotros forestales han sido demonizados, y FSC tiene el aprobacion de los ecologistas, ahora podemos decir que nosotros somos tambien ecologistas*" [Because we the loggers have been demonized, and FSC has the approval of the environmentalists, now we can say that we are environmentalists too.] I've heard this kind of statement from many timber company employees explaining their interest in the FSC, but all were officials or scientists with a clear investment in their company's green reputation. It's interesting now to see the field staff at Rio Largo solemnly nodding their agreement.

At this point Luisa turns again to her audience—how are *they* going to commit to making this work? And she goes through them one by one, asking for a specific commitment in terms of the job that they actually do. Most of them, a little uncomfortably, describe ways of 'doing their job right'; being careful and using the safety equipment, keeping track of the property and how the trees are growing, not throwing garbage out of the truck, and so on. Watching this conversation emerge, watching loggers struggle to articulate their commitment to an abstract entity with a name they can't pronounce, I can't help feeling that there's a tautology at work. Luisa has used "sustainable" management interchangeably with "responsible" management throughout the presentation, clearly trying to take the sting out of the alien term by linking it firmly to a familiar one. Then to close the presentation, she is essentially asking the workers how they plan to be responsible so that they respond by saying they'll be responsible. I'll do a good job by doing a good job. I don't necessarily sense that they've developed an idea of FSC or 'responsible forestry' as an *ecological* practice, or of sustainability as something categorically different than what they already do. And Luisa makes an effort to confirm this. She congratulates the various commitments, and says that this is how to make FSC into something "*superconcreta.*" Perhaps before the training, she suggests, the staff had only a vague sense of what their FSC certification meant. She playacts: "*¡FSC es como, wow! No sé, es una cosa gringo, pero no! Cosas practicas!*" [FSC is like, wow! I don't know, it's a gringo thing, but no! Practical things!]

This chapter is concerned with those "practical things," and how they relate to the "gringo things" and the ideas of the "big thinkers." In the previous chapter I introduced the concept of interpellation, the production of subjects through discursive authority, using Althusser's metaphor of the policeman shouting "Hey you!" in a public place. Here I would like to add that Althusser's metaphor of interpellation uses a policeman for a reason. If the 'hailing' came from, for example, a neighbor, I am also free *not* to recognize myself, not to look up. The cop, on the other hand, can *make* me look. The disciplinary capacity of the authoritative discourse, while it tends to operate through interpellation of subjects, does not depend entirely on the subject's agency. In the case of forest certification, as I've noted, any number of subjects respond to the hailing that comes from neoliberal environmental 'regulation,' some of them with greater access to discursive and material resources than others. The narrative about training forest workers provides an entry to theorizing about indigenous knowledge in relation to the politics of representation for both indigenous and workers' rights. As I hope will become clear, in all cases the knowledge of less powerful actors in FSC-Chile appears as a representation. This is true of the knowledge of forest workers and of the indigenous knowledge of FSC-Chile's lone Mapuche representative. But as Luisa's training suggests, there's a big difference between the people sitting at tables to negotiate the standards and the people in and around forests who will live and work with the consequences. What's more, when we speak of discourse and interpellation we risk lapsing into the kind of textual bias that cultural studies are generally accused of harboring. In this chapter I will consider how the discursive aspect of knowledge, as represented and codified by FSC standards, is troubled by other ways of knowing—what we might call 'knowing from below.' At first glance this seems to imply a space outside the standards' textual field, but I propose to think about FSC standards across different ways of knowing them.

INTIMATE KNOWLEDGE

When "knowledge" and "nature" come together, the anthropological imagination leaps to a construction we've come to know as "environmental knowledge," "traditional ecological knowledge," or "indigenous knowledge." The anthropological study of indigenous environmental knowledge has deep roots—in particular the disciplinary transition when economic botany (exploration of the tropics for useful plant products like rubber and rattan) began to shade into ethnobotany—research informed by the comprehensive botanical knowledge of local people. This transition is often embodied in the totemic figure of Harvard ethnobotanist Richard Evans-Schultes (see Evans-Schultes 1995). Several of Evans-Schultes's students went on to become important figures in contemporary ethnobotany—Wade Davis (see *Serpent and the Rainbow, One River*, and others) and Mark Plotkin (*Tales of a Shaman's Apprentice*, founder of Shaman Pharmaceuticals) being the best known.

At about the same time (1950s–1960s), the merging of field ethnobotany with cognitive anthropology was producing important synthetic work that came to be known as *ethnobiology*. This subdiscipline described the development of taxonomic knowledge in a universalist-evolutionary format (Berlin et al. 1966; Brown 1977; also see Berlin 1990). In short, the argument is that subsistence strategies correlate closely with the complexity of taxonomic knowledge—'simple' foraging societies have thin taxonomic trees, while 'complex' agrarian societies identify and classify a much more complex range of species. Ethnobiologists of this period argued that species categories were defined by evident disjunctions existing in nature, and as societies developed in social complexity they developed the means to identify and name those self-evident categories in more and more detail.

The evolutionary reductionism implicit in much of ethnobiology did not go uncontested—the debate between this cognitive universalism and a utilitarian/adaptationist position marks a defining moment in the field. Eugene Hunn's work with Native American tribes in the Pacific Northwest documents taxonomic strategies that don't adhere to the monolineal model at all. He argues that people recognize a 'natural core' of species according to evident perceptual cues, but that most taxomic categories outside that core are based on interactive constructions of salience defined by actual uses of plant and animal species. Examples include Sahaptin names for roots crops (all known as *xnit*, for 'something you dig') or Maya butterfly taxonomies (very few taxa for adult butterflies, despite very evident differences in color and size, but many taxa for larvae, according to their being edible, toxic, or harmful to crops [Hunn 1982]). This distinction made a crucial bridge in the field between *knowledge* and *behavior*, left open in the supposedly evolutionary models of the earlier ethnobiologists.

Working with the same ethnographic record, Timothy Ingold develops a dramatically different take on environmental knowledge (Ingold 2000). Even those ethnoscientific frameworks that explicitly link knowledge and behavior maintain an epistemic divide. Not, in Ingold's argument, between 'scientific' and 'indigenous' knowledge; but rather between the knower and the known, distinct actors on a spectrum from the 'social' to the 'ecological.' A conventional model of indigenous knowledge suggests that an indigenous hunter-gatherer has a lexicon of knowledge about a given natural environment or set of species that he then enacts in the process of procuring resources. By contrast, Ingold offers a phenomenological approach: The hunter (or gatherer, or scientist) creates the world in the specific moment of interaction with the prey animal (or plant, or landscape, or . . .). The set of facts that the hunter can articulate verbally to an interviewer is a schematic representation of a holistic set of multi-sensory and interspecies interactions. Ingold rejects dichotomous models for these processes, as in Rappaport's 'cognized' and 'operational' models (Rappaport 1968), in favor of a philosophy of dwelling and enskillment. What we see as knowledge about nature is best understood as a whole-body process incorporating the senses, muscle-memory, and even kinesthesia (Ingold 2004) alongside verbally articulated knowledge. What Ingold calls dwelling, or being-in-the-landscape, is a sensuous as well as intellectual process.

Hugh Raffles also engages the slippery contexts of thought and knowing in which knowledge of nature is generated. In the paper that provides the title of this section, he considers the settings and relationships in which and through which rural Amazonians come to know their rivers:

> how on these rivers people enter into relationships among themselves and with nature through embodied practice; how it is through these relationships that they come to know nature and each other; and how the relationships, the knowledge, and the practice are always mediated not only by power and discourse, but by affect. And it also brings to

mind how affect, though inconstant, is also ubiquitous, the perpetual mediator of rationality. (Raffles 2002b, 326)

Raffles describes the affect and relationality that condition knowledge as predicates of *intimacy*; intimacy between people, intimacy between people and the landscape. He suggests as well that the inability to recognize and respect that intimacy, and the vulnerabilities that it implies, can contribute to the sort of problems that natural resource managers often encounter across cultural difference, "in circumstances of confounding asymmetry, circumstances in which the success of a project depends on recognition, sensitivity, and the willingness to face the often painful complications of relationality" (2002b, 326).

So we begin to see how actors at different levels of the hierarchy in the Chilean forest sector might come to the space of certified forest management not only with distinct educations, taxonomies, and skillsets, but also with distinct embodiments, sensations, and affects. As Raffles suggests, these affects and sensations are conditioned by the histories and power dynamics in which management regimes take shape. What kinds of rationality and relationality will emerge?

PRINCIPLE 4

Perhaps the definitions and practices of ecological sustainability, as imagined in the FSC standards, do not appear in the experiential worlds of the forest workers. The workers, however, do appear in the text of the standards. Consider the text from FSC-International's 'generic' Principles and Criteria:

PRINCIPLE 4: COMMUNITY RELATIONS AND WORKERS' RIGHTS

Forest management operations shall maintain or enhance the long-term social and economic well-being of forest workers and local communities.

Criteria:

4.1 The communities within, or adjacent to, the forest management area should be given opportunities for employment, training, and other services.

4.2 Forest management should meet or exceed all applicable laws and/or regulations covering health and safety of employees and their families.

4.3 The rights of workers to organize and voluntarily negotiate with their employers shall be guaranteed as outlined in Conventions 87 and 98 of the International Labour Organisation (ILO).

> 4.4 Management planning and operations shall incorporate the results of evaluations of social impact. Consultations shall be maintained with people and groups (both men and women) directly affected by management operations.
> 4.5 Appropriate mechanisms shall be employed for resolving grievances and for providing fair compensation in the case of loss or damage affecting the legal or customary rights, property, resources, or livelihoods of local peoples. Measures shall be taken to avoid such loss or damage.

It is important for me to note as well that many of the leading officials in FSC-Chile, while technically trained in forestry and formally assigned to the Environmental chamber, considered themselves to be workers' advocates. Luis Astorga, current president of FSC-Chile, has a long history of supporting timber workers' unions and participating in outreach and advocacy. In terms of his role in supporting workers' participation in standard-making, consider this conversation:

LA: Recently it's occurred to me that I, there's two fundamental aspects that the social component in the FSC must have. One aspect is that it doesn't consider the social component as a system, as part of a system. It considers it only as dispersed actions. Now, I act with the unions in one form. I act with the neighboring community in one form. I act with the indigenous communities in one form.

AH: Okay . . .

LA: Then there's not a system as such, rather there's a lot of subsystems, you might say. What's logical is that a company that wants social sustainability works the issue like a system of social planning. Just as there's a system of environmental planning, there also has to be a system of social planning, and there also has to be a system of economic planning. Economic planning the company already does perfectly well . . .

AH: That never fails.

LA: They've got no problem with that. Environmental planning, the companies have started to understand what they're dealing with and they're managing much better, especially when a large part of the companies are certified by ISO 14.000.[2] But the social planning has been dealt with in a dispersed form, in a paternalistic form, like saying, "We're the rich of the area, and the problem is a problem much further down." Because the system of social planning; the company first has to define its principles, to define its objectives as an issue, right? What's happening with all the people that I dealt with, with my workers, with my neighbors, with the communities that work with me, what objectives do I have for each of these groups? Then you have to define objectives, and the definition of objectives, of course, also brings up defining forms of fulfilling those objectives. For example, with the workers, there are at least two important aspects. One is the union aspect. A lot of companies oppose the formation of unions in Chile.

AH: Really? Oppose it altogether?

LA: They oppose it! Once a businessman that wanted to certify with FSC said to me, "I fail if a union forms in my company, I fail as a manager." He said that he failed as a manager because the union is not necessary. So it's clear here in Chile that there's a very strong disadvantage. . . . You haven't seen, for example, what happened a few days ago in Arauco. Yeah, in Arauco the problem was that the forest workers don't have unions. Or, no, each little company, there isn't a union. What worked with that was the federation of forest workers, which this federation did have certain forest unions of several companies. But finally all of these organizations were unified. Then the company, when it worked with each of them, switch to its contractors that didn't have a union . . . so in the end, that way the company did what it wanted and that's why now . . .

AH: Lots of companies work with contractors, though, right?

LA: Of course, of course. Then there was a problem where the timber company had to define concretely how to do its planning as far as its workers . . . Now, the system of contractors is not a sustainable system, at least until the power relations change, where the contractors have the ability to argue wages. If the company is only one and there are many contractors, a contractor that says, "No, I don't agree, okay?" Then they could get somewhere. But if someone tried that now, well, permanent unemployment. So, then they have to define a policy in relation to safety, to wages. We'll see how wages will improve, and what relation wages have with profits. If you read the Chilean press about forestry, I could tell you, for example, in the cellulose plants, they get higher prices every day, at least the companies have tremendous profits. Nonetheless, wages haven't gone up. If not for this decision, for this situation, and that worker died. You heard about this, in Curanilahue?

AH: No . . .

LA: Yes. Then that made the church and everyone really concerned, and the company had to give up a lot more than th[ey] had thought to give at the beginning. So, that's why . . . Well, but that wasn't a company certified by FSC. But if it was certified by FSC, we're proposing that there be an effective system of social planning. And that system of social planning, the company has to define aspects of its workers that are more sustainable than they currently are, right?

AH: Sure, there'd have to be a real change in the whole labor system.

LA: No doubt, there would have to be a huge change in the labor system. But the system we have is very weak in this respect.

I reproduce this conversation here in part to illustrate that Astorga as a leader in the FSC, along with other prominent voices in the FSC, does not conform to stereotyped expectations in terms of his relationship to forest workers. Anthropologists and others studying forest politics have documented the polarizing tendency between environmentalists and loggers in particular (Satterfield 2002; Braun 2002), but Astorga was not alone in his active concern for the rights of timber workers. At the same time, that concern served to reinforce the ascription of forest workers to particular spaces within forest certification.

At the meetings where FSC-Chile negotiated the particular terms and indicators to operationalize this general standard, forest workers were represented by a union official; a forestry engineer and plantation worker, at the time head of the syndicate of CONAF employees and now a higher official in CONAF itself. The board of FSC-Chile conducted three *consultas publicas* [public consultations], in Temuco, Valdivia, and Concepción. These were opportunities for any interested parties to evaluate the draft standards and the proposals for specifically Chilean indicators, to offer suggestions and criticisms. Invitations were sent to a somewhat undifferentiated group of possible attendees, with instructions to pass the invitation on to other interested parties. In the case of the Temuco consultation, at any rate, I was told that the invitation was simply posted outside of the CONAF offices. Those participating were assigned to one of the FSC's three chambers (Economic, Environmental, or Social) and split into working groups to address the relevant set of Principles and Criteria. Although this was not the intention of the board, the three consults divided fairly clearly along the lines of interested parties; the Valdivia consult was inclined toward representatives of the Environmental chamber, Temuco included many small forest proprietors that were assigned to the Economic chamber, and Concepción saw a large majority of representatives from the forest workers' unions. Perhaps as a result of this two-stage channeling process, the comments made by the various stakeholders are neatly divided according to their areas. That is to say, public commenters in the social chamber spoke up primarily during the Concepción *consulta*, and almost exclusively on Principle 4.

Here's how FSC-Chile chose to operationalize workers' protections: Criteria 4.2 states that "Forest management should meet or exceed all applicable laws and/or regulations covering health and safety of employees and their families." This is defined according to eight indicators, each with their corresponding measures for verification. Indicators 4.2.1 through 4.2.8 stipulate a written plan for the prevention of on-the-job accidents; records documenting the company's compliance with Chilean labor laws and terms of workers' contracts; documented availability of full safety equipment and training for all workers in how to use it; and evidence that workplaces (planting and harvest sites, sawmills, etc.) are maintained in safe conditions. Criteria 4.3 states that "The rights of workers to organize and voluntarily negotiate with their employers shall be guaranteed as outlined in Conventions 87 and 98 of the International Labour Organisation (ILO)." This is a major concession on the part of timber companies, as Chile in general does not stipulate this level of protection for workers' rights to organize. Indicators 4.3.1 through 4.3.5 stipulate that companies will not infringe on the rights laid out in the ILO Convention (although these are not explained), and that information about their rights to organize will be available to all workers. In all cases, the verification of compliance consists of "consultation with interest groups."

There are two issues that appear to emerge from the articulation of workers' interests into FSC standards. First, built into the method and process of standard-making are a set of assumptions about what issues forest workers are able to speak to. Safety, wages, the right to organize; these are the concerns of forest workers. At no point, however, did the framework allow for the possibility that workers or their representatives might have valuable information to contribute to, say, Principle 6 (environmental impacts) or Principle 8 (monitoring and evaluation). Second, forest workers in each case are represented *for*. Timber companies assume the responsibility for providing information about workers' rights, for ensuring their safety, for consulting with workers or their representatives. "Consultation with interest groups," in Chile and most likely elsewhere, is a famously toothless requirement (see Angell and Graham 1995; Irwin 2006). Consultation, after all, can mean many things—the *capacitación* that opened this chapter represents an episode of consulting with forest workers, yet they clearly were not contributing to Rio Largo' management plan or means of engaging with certification. Consultation does not provide the consulted party with any veto power, any meaningful authority, or even necessarily any input at all. In both of these senses, the resulting standard reinforces the paternalistic model to which Astorga alludes.

The forest workers are better represented and have their rights much better taken care of than the Mapuche. They still pay the costs of representation, though—they are spoken for, not speaking. There is some factor at work in the process by which FSC standards are set that places these more intimate ways of knowing outside the sphere of influence. Despite significant differences, forest workers and Mapuche share a subaltern predicament that prevents either subject position from being actively realized within the project of certification. Critical ethnographies have demonstrated how other institutions of accountability generate virtual versions of the collectives (firms, customers, students, citizens) on whose behalf they conduct their audits and reviews (Miller 2003; Strathern 2000; Power 1997). The gap that results between the represented and the representation is a generative one, allowing the efficient mobilization of those collectives for purposes that may or may not map onto their own interests. Indigenous knowledge more than many ways of knowing has been subject to representation and radical reinterpretation (Brosius 1997); its relation to standards in particular leads me to consider alternative frameworks for considering these knowledge forms in the context of governance regimes.

THE SMELL OF THE LAUREL

During an earlier visit to Temuco, capital of the IX Region and center of Mapuche radicalism, an activist had encouraged me to look up a *werken* (community spokesperson) who I will refer to as Gonzalo. Completely surrounded by pine plantations, Gonzalo's small community had been a hotspot of Mapuche resistance. The situation seemed particularly pertinent to my research, as the company enclosing Gonzalo's community had been bought by a larger company that went on to obtain FSC certification on other properties. What did that mean for this FSC business, the first activist wondered, if a company responsible for oppressing indigenous communities in one place could get the green seal of approval for responsible forestry in another place? It seemed like a good point to me, and I did my best to track down Gonzalo. Months later, when we were finally able to get together for *mate* and forestry talk, the issues had all moved on. Gonzalo's family had moved, the timber company had divested some of the properties, and some members of the resisting community had ended up contracting with the timber company as seasonal workers. Lacking a conflict to expound on, we sipped *mate* and talked about mutual acquaintances, German tourists, and weather. As I was leaving, though, I off-handedly mentioned that I would soon be meeting with Mapuche forester Pablo Huaiquilao (see the following), who had some interesting ideas about *conocimiento indígena* [indigenous knowledge].

Gonzalo seemed startled. Why hadn't I mentioned that I was interested in indigenous knowledge? He was an expert! He seemed genuinely baffled that I'd asked so many questions about timber companies if indigenous knowledge was what I was after—and wasn't I an anthropologist, anyway?[3] He immediately began rummaging in the back room for evidence of his expertise, so I risked missing my bus to see what he would produce: Dried herbs, arcane ritual devices, photos of foraging in remote forests? Emerging from the back room, Gonzalo handed me a Mexican textbook on medicinal plants; he was taking a class at a nearby cultural center, taught by a university student, funded by German aid money.

I share this anecdote to point out that my argument about practice and situating knowledges cuts both ways. Just as the application of scientific forestry becomes contingent and culturally bound in its practice, the thing anthropologists and development workers call 'indigenous' or 'traditional knowledge' turns out to be more contemporary, local, and integrated than the name would suggest. Encounters like Gonzalo's, like many appearing in the literature (Ford 1999; Sillitoe 1998; Baviskar 2000), point out a tension between facts, texts, and indexical knowledge on the one hand, and the unspoken, unwritten qualities of embodiment or practice on the other. This kind of tension, I argue, is an essential quality of this thing we call indigenous knowledge.

The study of indigenous knowledge in Chile is limited, for the most part, to health and medical issues (Duran et al. 2000; Morales 2000). Mapuche medicine, in particular, has seen a sort of popular resurgence, visible in the form of Mapuche pharmacies in urban areas, offering herbal medicines prepared by state-certified "bicultural practitioners." When it comes to forestry and the fierce politics behind it, indigenous knowledge has a much lower profile. In fact, I would say it's marked largely by its absence. Agricultural and forestry planners alike describe Mapuche *campesinos* as poor farmers and natural resource managers, by virtue of their poverty and lack of education (Ladio and Lozada 2004). The suggestion of knowledgeable or effective indigenous management strategies is often dismissed on the grounds that any meaningful tradition of indigenous knowledge would have vanished in the process of acculturation and assimilation (Saavedra 2000). The rhetoric of indigenous knowledge in development, nearly hegemonic on the global scene (Nygren 1999) has yet to gain more than a slight foothold in Chilean forestry. In a context like the negotiation of FSC standards, where the principles are predicated in part on global discourses of both indigenous rights and indigenous knowledge, this lacuna can raise real complications. I discussed this at some length with Pablo Huaiquilao:

A: Can you give me an example, something more specific?
P: I'll give you an example, yeah, a really concrete example of a discussion that we had that was just like that. We'd had discussions all day, and in the afternoon we were talking about intellectual property rights for traditional knowledge, how to compensate it. So, uh, well, the people were tired, but it had to be discussed either way, so I raised my hand and definitely, the people, especially the people from the timber companies, they got mad. And I had a moment, a really tough, really uncomfortable, really, how should I say, very, uh, aggressive on the part of them against me. Because they said definitively that there was no knowledge that the communities or the Mapuche world could contribute to forest management. And that, well . . . I reacted too in a really tough way with that type of statement. So there we had a really tense moment within the negotiations. Really, really tense.
A: Wow. Can you explain for me what type of knowledge might be relevant in such a case?

P: See, in general I think that you have to look at the use the Mapuche world has traditionally made of the trees. In general, of nature. Uses like, above all medicinal. And that the timber companies do not want to recognize.

A: And is that because they just don't deal with medicinal things, they're not concerned, or because . . . ?

P: No, what happens is that the timber companies come to a forest for nothing more than the value of the wood. So, if we're going to construct a standard, we have to see all the cases. If the company, I don't know, finds a traditional use that's attractive for their market, we have to have safeguards, within certification, so that this traditional knowledge will also be recognized as not theirs, but rather a traditional use.

A: Okay, right. So, for example, a company discovered a particular plant on their property and wanted to bring it to market, it couldn't be . . .

P: Right, that type of thing. This was a really, really controversial subject at the time.

A: Of course. Worldwide.

P: And at the time, everyone just wanted . . . that there wouldn't be any type of contribution that the communities could eventually make, the indigenous communities. And it should just be, "yeah, approved, approved, let's go home."

Recognizing the significance of indigenous knowledge about forests is a material barrier to agreement between timber companies and social interests in the FSC, one that environmental interests are not well-placed to bridge. Most important, though, is the fact that the thorny issue of intellectual property is central to the recognition of indigenous knowledge in the first place, for timber companies and the indigenous advocate alike.

When social scientists use terms like 'indigenous knowledge,' 'traditional ecological knowledge,' 'local knowledge,' or some acronymic version thereof, we imagine that we are referring to a fairly recognizable object of inquiry. As the term leaves the realm of our disciplinary debates and enters into wider circulation, however, it becomes subject to contestation from multiple positions. A number of actors are invested in some version of indigenous knowledge—academics, conservationists, and intergovernmental bodies as well as NGO advocates and indigenous people themselves. As these actors strive to collaborate on joint projects, from community forestry programs to the Convention on Biological Diversity, their divergent definitions of the term often lead them to speak past each other. What do people mean when they say 'indigenous knowledge'? To ask the question is to enquire into complex and power-laden histories.

A diverse assortment of anthropologists and fellow travelers has created a rich body of literature on non-Western sciences and technologies—in the conservation and development context often referred to as Traditional Ecological Knowledge (TEK). The theoretical perspectives in this literature have undergone considerable transformations in response to new data, new trends in social theory, and new sociopolitical conditions. Emerging from and parallel to the academic ethnobiology discussed above, applied anthropology, rural sociology, and related fields began in the mid-1970s to address the issue of indigenous knowledge in practical development and conservation contexts. The ethnographic data and epistemological perspectives on indigenous knowledge drawn from academic anthropology resonated with new methodologies for development, including participatory rural appraisal (PRA) and participatory action research (PAR) (See Chambers et al. 1989; Moock and Rhoades 1992; see Kindon et al. 2007). Advocates of indigenous knowledge argued that development interventions that failed to take into account local knowledge and local management practices were both unjust and doomed to fail. Within conservation research, indigenous knowledge emerged as an important aspect of the field via studies of common property regimes and wildlife management (Berkes et al. 2000; Posey 1985; see Drew and Henne 2006 for an overview).

In the context of Third World development, the polemic involving indigenous knowledge served a valuable purpose, defending the knowledge and capacity of the local against what was seen as a hegemonic global scientific machine. At the same time, critics have situated indigenous knowledge in a theoretical space that is much more constructivist; many question the epistemological basis for distinguishing between indigenous and scientific knowledges at all (Agrawal 1995; Ellen and Harris 2000; Turnbull 2000; Watson-Verran and Turnbull 2001). Others address the way ideas about indigenous knowledge are shaped by articulation with conservation or development discourses (Hobart 1993; Richards 1993; Nygren 1999), or the implications they hold for ethnic identity politics (Li 2000).

The constructivist approach to indigenous knowledge raises some concerns about the nature of indigenous knowledge itself as an object of study. One might be reminded of Latour's uncharitable comment about the social sciences: "The expression would be excellent except for two drawbacks, namely the word 'social' and the word 'science'" (2005). Likewise, the term 'indigenous knowledge' carries its own internal contradictions. Referring to the indigenous (or its corollaries, the local or the traditional) implies a topology, a spatial limit to the knowledge in question that may not hold up to analysis. 'Knowledge,' in turn, is a blanket concept that conflates information, cosmology, and management practices, while leaving a number of assumptions unexamined.

Properly *locating* indigenous knowledge is a problem that has troubled the field throughout its history, from colonial exploration to ethnobiology to community-based conservation. Like much of anthropology in general (Gupta and Ferguson 1992; Pigg 1992), the study of indigenous knowledge depends on the persistent association between place and knowledge, between locale and the local. Our tendency to distinguish between indigenous and

Putting Knowledge to Work 71

scientific knowledge constructs indigenous knowledge as a holistic single entity mapped directly onto a bounded cultural group. Indigenous people are the holders of indigenous knowledge, which pertains to natural entities within indigenous territory; the spatial logic is circular.

It is no longer novel to suggest that indigenous knowledge is more contemporary, integrated, theoretical, and even global than the name might imply (e.g. Agrawal 2002; Brosius 1997, 2001; Dove 2002; Ellen et al. 2000; Ellis and West 2004; Escobar 2001; Hayden 2003; Li 2000; Nadasdy 1999; Nygren 1999; etc.). After all, more than a decade has passed since Agrawal's (1995) incisive deconstruction of 'indigenous' and 'scientific' knowledge categories. More sophisticated perspectives on locality and epistemology are beginning to make themselves felt in environmental policy discussions (Brosius 2006). But the constructivist approach to indigenous knowledge has in no way replaced or eliminated the older romantic or evolutionary perspectives. Feeling, as I do, that the epistemological basis for distinguishing indigenous from scientific knowledge is shaky at best, how are we to make sense of the concept's persistence in academia, in conservation and development, in the hands of indigenous farmers and foragers themselves? As Li notes, "If such knowledge exists everywhere, in city and country, west and east, then the distinctive feature of 'indigenous environmental knowledge' is not its content but rather its *location* in particular agendas" (Li 2000, 121). Indigenous knowledge in one form or another occupies a place in a variety of international agendas: Sustainable development (Sillitoe 1998; Warren et al. 1995), human rights (Maybury-Lewis 1992; Nygren 1999), and intellectual property rights (Brush 1993; Dutfield 2002; Moran et al. 2001).

Locating indigenous knowledge within this shifting assemblage is not just a matter of identifying actors in a political debate, creating a typology, or naming stakeholders. Rather, we need to trace the relationships between indigenous people, their advocates, governments, academics, corporations, and NGOs in action, as they emerge and define the networks by which knowledge circulates and accrues new meanings. This suggests that, as Nygren argues, "the focus of analysis should be on those processes that legitimize certain hierarchies of knowledge and power between local and global (scientific) knowledges" (1999, 268). The processes in question play out in a number of venues: From social science classrooms to the offices of international environmental NGOs; from sustainable development project sites to ecological field research sites; from World Bank meetings to the World Parks Congress. Such a development might represent a stretch for researchers in the social sciences, to the extent that we feel confined to our disciplines or applied or policy-relevant frames, but our informants and interlocutors may be out ahead of us in this respect. Consider the conversation that emerged when I asked Huaiquilao how his forestry career began:

> P: See, what happened is that when I was little something very strange happened to me. I always dreamed about a particular smell,

I smelled this particular aroma in my dream, that I'd never encountered, that I'd never experienced in my real life.
A: Interesting.
P: And I had another recurring dream that, I saw myself in a forest. I mean, I didn't see myself, but I knew that I was moving through a forest, but I didn't know what I was, because I couldn't see myself. I saw my way through the forest, I climbed some trunks, came back down, and moved really easily through the forest. I was an animal, see, because the dream was very precise, I passed very freely through the forest. But I never saw what I looked like.
A: That's also pretty interesting.
P: And the subject of that aroma that I smelled in my dream was very . . . it really touched me. Then, apart from all that, I really liked nature. In the 8th Region there wasn't much left, then I went to study in Temuco, at Universidad de la Frontera.
A: And at that time, in the region, were there a lot of plantations?
P: So many. I grew up among pine plantations, I played in pine plantations, I played on the pine trees, I climbed the trees, eh, but I did feel the need to find other things, that I had some kind of internal connection without really knowing about. Then when I came, I knew that I really liked nature. And so I chose ingeniero forestal. And the first field trip that we had was to a native forest, near the University, very impressive. And when I entered the forest, right, walking through the forest, impressed, and right away my heart started racing, and I couldn't figure out why, and I started to analyze myself and see what was happening to me, because there it was. I was very excited. And all of a sudden I realized that I was smelling the aroma that all my life I had smelled in my dreams. And it was the smell of the laurel. The laurel has a very distinct smell, spicy. And of course, I get closer to the smell, and I take the branch and identify it absolutely, that it was the smell from my dreams. Then it was very, very strong. And there I knew that this was my future, that it was exactly the path that I would always . . .
A: It's a sign.
P: Yes, I believe it was. No, more like . . . a return to my origins.
A: OK.
P: A return to my origins, in some way my spirit called me back to connect me to the native forest.
A: At that point, had you talked about this stuff with your parents?
P: You know, no, no, I didn't let it out. It was . . . I don't know, I felt that it was something really strange, but that I didn't have any better explanation. Then while it was growing and I was in university, there I started to express it. And, well, there in the university, eh,

Putting Knowledge to Work 73

I realized a lot of things. The truth is we're immersed, I realized, eh, for example, the harms of the timber plantations, not only in the Mapuche communities, but to the life of the whole region.

A: Was that the first time that you perceived that, the impact of the plantations?

P: I think that when I was a boy I also realized. Or maybe like a lot of things that I felt at the time, but no, like I was more mature before I internalized it. And when I came to the university I realized a lot of things, and there began a really strong struggle with the things that formed me professionally.

A: This was in the 1990s?

P: Yes, in the 90s, in university. The university had then, I don't know about now, but then, it was a very technical training, and directed too towards managing plantations.

A: Of course, oriented most toward production. Which professors did you study with?

P: Fernando Hartwig . . .

A: Really?[4]

P: Who was . . . for many years he was chief of production for Forestal Mininco. And yeah, also with the other, the other side of forest management, which for me, which served me much more: Patricio Nuñez, Ricardo Molina, basically. For years he was chief of production for CONAF.

A: Oh, I've heard a lot about him, yeah.

P: Yeah. And well, but what brought me most to management . . . was finally, the knowledge of the problems that had been caused by timber plantations.

A: And that was a controversial subject in the school?

P: In fact I was identified very clearly [laughs] as one of the most controversial students in the silviculture classes, you know? In fact I, uh, did a lot of extra-curricular activities in order to, uh, to make the students question a little bit the type of training that they had within the university. Then the professors had me really [laughs] . . . sometimes really isolated, sometimes really integrated.

A: Where there resources for this subject, to pursue the idea within the university?

P: Yes, yes. Fortunately yes. Well, small projects within the university, small projects to bring, for example, *kimches*, who are Mapuche 'wisemen,' to do workshops . . .

A: OK, that word was . . . ?

P: *Kimche*. To make, uh, workshops about Mapuche cosmovision and the relation to the environment or to the native forest, we made activities a few times. I also got a project to bring young middle-school students to the native forest, to have them for a

74 *Putting Knowledge to Work*

whole day immersed in the native forest, and to be able to really soak them in the love of nature. And also another project that . . . another project to bring students from the university in general to certain national parks. Then, well, fortunately there were resources for these things. In addition to work that we had with some of the Mapuche youth organizations to look at the whole problematic of the timber plantations and the consequences in Mapuche property.

A: Alright. And was there a lot of interest in these activities?

P: Yes, lots, lots. It was a really strong subject. And we were able to bring the subject up to some important Mapuche leaders, although not to the directors or whatever of the companies, that one way or another . . .

A: Did this larger political struggle cause problems among the students in the forestry program? Where relations difficult with the more plantation-oriented students?

P: Yes, yes. Look, in the university they sell an image of, of the successful ingeniero, uh, that works for a big timber company. It was selling a lot at the time . . . , now I. Now I hope not. But at the time it was selling a lot. A lot of my classmates dreamed about driving around in a 4x4 on a timber plantation. You know? So that was selling a lot at the time. And it was, I think, together with certain professors that we were able to a get those same students to question a little bit many of the things they were taught.

Note how Pablo's experience with the mysterious dream-aroma takes us back full-circle to Ingold's argument that knowledge emerges in full bodily engagement with the non-human world, at the earliest pre-discursive level. But in Pablo's account we hear something more; a seamless transition from the sensuous and mystical to the political. Hearing this story, I must remind the reader that Pablo is an activist and an *ingeniero forestal*; not otherwise given to talk of magic or spirituality, he is as practical and focused on the 'concrete' as Luisa's forest workers. Understanding Pablo's activism and the knowledge politics behind it requires a new framework in which to consider indigenous knowledge. Ethnobiology, where "knowledge" implies a lexicon of facts to be known, offers little purchase for epistemic politics or the subjective experience of someone like Pablo. If we think of indigenous knowledge as an aspect of global development discourse (Nygren 1999), we lose the qualities that make Pablo's dream meaningful, along with all of the lively landscape actors in Ingold's phenomenological perspective (2000). It's not clear, though, how Ingold's epistemology of dwelling works with the politics of indigenous knowledge that inevitably accompany them wherever they articulate with governance and management regimes. For Pablo, we see no clear separation between Mapuche religious traditions, medicinal practices, his youthful dreamlife, and the politics of expanding timber plantations in Mapuche territory; and he enacts this free set of associations in his forestry extension work and educational workshop activities as well. What kind of thinking about indigenous knowledge will let us keep up with Pablo, and keep all of these problematics in focus at one time?

KNOWLEDGE, REASON, AND IMAGINATION

I would argue that we need a new framework in which tradition, imagination, indigenous identity, territory, intellectual property, and the global reach of conservation and development all find space. To begin with, if Agrawal (1995) and other critics are correct, we cannot continue to treat indigenous knowledge as a *sui generis* category in isolation from other ways of knowing the natural world, or in isolation from its politics. In addition to the failure of epistemic clarity that Agrawal demonstrates, isolating indigenous knowledge constrains our ability to understand local articulations with governance and management regimes; to understand what standards mean for forest workers and how forest workers apply and reinterpret standards. A better perspective might imagine the situated knowledges of less-powerful actors (indigenous and otherwise) as co-constitutive of more widely circulating forms like FSC standards. At the least, such a perspective would be an appropriate response to the postcolonial call to put the local and the global in the same frame.[5]

If the existing frameworks for knowledge as facts, phenomena, or discourse share a pattern of overlapping gaps, it would seem appropriate to look for common ground or space of agreement between them. In one of the best synthetic arguments on indigenous and local knowledge, Ellen and Harris argue:

> [that] a baseline of universal reason exists and that in all traditions it is driven by shared human economic needs and cognitive processes, but also that they are activated and expressed in different cultural contexts ... We believe that IK [indigenous knowledge], in the sense of tacit, intuitive, experiential, informal, uncodified knowledge, will always be necessary and will always be generated, since, however much we come to rely on literate knowledge which has authority, the validation of technical experts and is systematically available, there will always be an interface between this kind of expert knowledge and real-world situations. It will always have to be translated and adapted to local situations and will still depend on what individuals know and reconfigure culturally independently of formal and book knowledge. (Ellen and Harris 2000, 28)

The argument for a universal human reason in the context of indigeneity and environmental knowledge is a fascinating one. While it echoes the 'psychic unity' project of cognitive anthropologists and ethnobiologists (Brown 1977; Berlin 1992, etc.), for Ellen and Harris the universality seems to exist at a more basic level—not in the taxonomies, or in the technique of taxonomizing (to coin a phrase), but in the "economic needs and cognitive processes" themselves. Certainly it is in this realm of economic needs and cognitive processes, or to put it differently, ways of valuing and ways of knowing, that projects like the Forest Stewardship Council propose to operate.

Finding human commonalities at a material level, beneath a framework of *cultural* difference and even relativism, has achieved the status of common sense in most anthropology and development research. This proposed baseline of universality has recently been challenged, despite its modernist appeal, by a (re)emergent anthropological attention to otherness shaped by the ontological turn in contemporary theory; what Stengers (2010) or Latour (2011) call *cosmopolitics*. Within anthropology, the ontological perspective in a sense forces our long disciplinary tradition of cultural relativism to its logical conclusion; what if the divergent beliefs of an Other, in spirits or shapeshifters or earth-beings, are true? Not just true-in-context, but true on an equal footing with our own, in defiance of any common ground in laws of nature or physics. Notable ethnographic examples include Eduardo Kohn exploring the dream life of dogs among the Shuar (Kohn 2007) and Nadasdy's reflections on a rabbit that, following Kluane First Nation tradition, "gave" itself up in his trap (Nadasdy 2007). We are forced to move from multiculturalism to multinaturalism (Vivieros de Castro 2005; Latour 2011), and evolutionary ethnobiology or cognitive modeling become notably irrelevant. Blaser (2010) develops a position for anthropological theorizing about globalization out of and through a Paraguayan Yshiro ontology, while Escobar (2008) considers the implications of such indigenous and subaltern philosophical otherness for social movements and alter-globalization. De la Cadena in particular notes the challenge that the more-than-human entities of indigenous ontologies can present to conventional notions of what constitutes politics:

> The appearance of earth-beings in social protests may evince a moment of rupture of modern politics and an emergent indigeneity. . . . an insurgence of indigenous forces and practices with the capacity to significantly disrupt prevalent political formations, and reshuffle hegemonic antagonisms, first and foremost by rendering illegitimate (and thus denaturalizing) the exclusion of indigenous practices from nation-state institutions. (de la Cadena 2010, 336)

For Escobar (2010), the conjuring of these earth-beings into politics represents part of a widespread trend in Latin America towards incorporating the full radical potential of culture and social movements across the spectrum of political action.

Some of these innovative reframings of indigenous ontologies are more attuned to historical and political dynamics than others, as Starn notes in a critique of Viveiro de Castro's influential model of indigenous perspectivism (Starn 2011). Like the opposite effort to locate a universal plane of reason beneath difference, though, even those accounts for whom indigenous ontology provides a lens for understanding globalization (Blaser 2010; Escobar 2008) still leave open the postcolonial challenge to place 'local' and 'global' forms in the same field. Answering that call requires considering historical encounters (including colonialism and globalization, but also development, revolution, and others that exceed our categories) as constitutive of reason at both ends of the spectrum. What we sometimes refer to as Western science is inseparable from indigenous knowledge not because both rest on a bed of universal reason, but because each has developed in historical relationship with the other. Anthropological research in two postcolonial contexts provides evidence for considering the historical coproduction of knowledge forms in this light: K. Sivaramakrishnan's study of scientific forestry in late colonial India, and Michael Dove's work on rubber management in Malaysia and South America.

For Sivaramakrishnan (2000), forestry serves as a site of colonial knowledge production, inasmuch as scientific forestry gathers together information, ideology, and practice. In spite of the arguments of some postcolonial critics (i.e. Shiva 1988), it would be inaccurate to say that mobile, global, and disinterested European science was imposed wholesale on Indian peoples and landscapes. Certainly colonial planners attempted to define modular planning regimes according to top-down principles often delivered directly from England, but social and ecological contingencies kept forcing them to be more and more particular. "Local knowledge," that is, knowledge of Indian locals rather than British expatriates, in Sivaramakrishnan's argument is not a knowledge system or alternative cosmovision, but rather a set of practices and acquaintances with specific places:

> Such localized production of information, and the processes of translating it into standard terms . . . made possible a project like scientific forestry" (63). In other words, it's a productive tension, productive of what we now know as 'scientific forestry'. Empire couldn't have proceeded otherwise, without enrolling the participation of locals—"neat division could not always be sustained in practice: cultural operations were often crucially dependent on local ecological knowledges and understanding the social mechanisms by which scarce labour could be secured for silvicultural tasks. (64)

In short, the scientific knowledge of expatriate foresters and the local knowledge of their field workers cannot be considered in isolation from each other or from the colonial regime that depended on and defined their interactions.

Dove (2000) documents the hybrid history of rubber silviculture as it traveled from South America to Southeast Asia. Notably, *Hevea brasiliensis* came to SE Asia via the Royal Botanical Gardens at Kew, based on seeds smuggled out of Brazil by an English cultivator named Henry Wickham. Wickham also attempted to export a large plantation-based cultivation system like that practiced in Brazil, but once transplanted to colonial locations in Indonesia and Malaysia, rubber cultivation went through several subsequent transformations. Although colonists attempted to establish plantations on the model Wickham advocated, a significant smallholder rubber sector developed alongside and often in competition with rubber estates. For smallholders, rubber was integrated into subsistence swiddening on a flexible, 'need-to-exploit' basis. As a result, planting and harvesting techniques began to vary dramatically from those practiced on estates, and incorporated practices developed for other crops. Cultivators in the estate sector denigrated smallholder practice and innovation out of competitive self-interest; state-sponsored research and policy initiatives tend to privilege plantations. In this context, Dove notes the problem of what kind of knowledge can be considered local: "The three discontinuities central to the historical production of rubber knowledge [are] separation from its origins, experimentation and changes in production technologies, and diversification of stake-holders . . . [all three are] inconsistent with the concept of indigenous knowledge, undermining its premises of localization, continuity and homogeneity" (Dove 2000, 234).

Taking the argument a step further, Dove emphasizes the dominant place of history in shaping this domain of knowledge, in the face of counterfactual possibilities—what might have happened had *hevea brasilensis* been brought to other colonies, or with a different freight of colonial management expectations along with it? Considering the contingent and historically situated qualities of all forms of knowledge in the rubber-growing sector, Dove cites Foucault's reference to knowledge production in the emergence of oppositions: "and what sort of concept of knowledge would produce (in opposition) a concept like that of indigenous knowledge?" (2000, 243). For Dove the question is rhetorical, but posing it seems inevitably to demand a postcolonial account of environmental knowledge that situates its 'locality' and historicity. As Dirks puts it,

> colonialism provided a theater for the Enlightenment project, the grand laboratory that linked discovery and reason. Science flourished in the eighteenth century nor merely because of the intense curiosity of individuals working in Europe, but because colonial expansion both necessitated and facilitated the active exercise of the scientific imagination. It was through discovery—the siting, surveying, mapping, naming, and ultimately possessing—of new regions that science itself could open new territories of conquest: cartography, geography, botany, and anthropology were all colonial enterprises. Even history and literature

could claim vital colonial connections, for it was through the study and narrativization of colonial others that Europe's history and culture could be celebrated as unique and triumphant . . . Colonialism therefore was less a process that began in the European metropole and expanded outward than it was a moment when new encounters within the world facilitated the formation of categories of metropole and colony in the first place. (Dirks 1992, 6)

Celia Lowe (2005) makes an argument for abandoning the ethnoscientific version of knowledge with its lexical baggage, in favor of an anthropological study of *reason*:

Although a study of knowledge can answer many of our questions about nature, it does not help us to see how nature becomes a question in the first place. A study of reason, on the other hand, operates as a metacommentary on knowledge. It shows us what will become valued as knowledge, how thought is actively structured and normed, and how these norms expand to cover a variety of situations. More than a hemispherically bounded analytic, reason can help to uncover the possibilities for, and conditions of, thought no matter where thought is located. (2005, 20)

Note that Lowe's reason differs from the universal reason proposed by Ellen and Harris, above, which continues to assume a discernable truth that can be isolated objectively—what Lowe calls 'knowledge.' Rather, she proposes a disarticulation of 'reasons of state,' critical reasoning, situated and partial 'knowledges from below,' and attempts to claim or project some form of universality: "Reason, as a folk category, entails those ideas, forms, and assemblages that attempt to erase their own conditions of possibility to stake universal claims, yet histories of thought as practice, and the locations from which claims on the universal are made, leave traces" (2005, 21).

Lowe's ethnographic study documents those traces across several fields of livelihood and biodiversity conservation in the Togean Islands of Indonesia. A particularly telling example appears in her narrative about local (Sama) shellfish divers. Sama divers pursuing deepwater pearl oysters and sea cucumbers often rely on two unique technologies: Portable air compressors mounted on boats that pump air down to divers at great depths; and a practice called *ilmu*, breathing techniques and certain prayers that allow divers to breathe underwater "like a baby breathes in its mother's womb" (2005, 118). By compressor or by *ilmu*, deep divers are able to harvest at depths and in quantities not possible by more normal means. Conservationists have expressed concern about the impacts of deep diving on coral reefs, and have attempted to eliminate the practice by banning the compressors. These efforts largely fail to connect, at least in part because conservationists cannot conceive of compressor diving as a practice closely related to the use of *ilmu*; to Indonesian scientists with university educations, *ilmu* cannot conceivably be other than backwards superstition.[6]

The smell of the laurel in Pablo's dreams led to a university education in scientific forestry, based on German models, neoliberal plantations, and post-Earth Summit forest conservation. He took that cosmopolitan (or cosmopolitical) body of thought into a career as a forestry extension agent, training small proprietors to work with native forests and incorporating the body of forest knowledge protected by generations of Mapuche. He also brought it to the table where standards for the certification of Chilean wood were negotiated, and on balance his contribution was rejected. Across the table, labor leaders represent and speak for forest workers. If their rights are better protected in the final standards, it is not because their experience of the forests in which they work has been validated. The stakeholder logic of the standard-making process leaves little room for a postcolonial imaginary, for a sense of reason that encompasses the phenomenological, the cognitive, the political, the historical. As anthropologists, we are not so bound; if we have tended to let categories of environmental knowledge be "fully formed objects that are simply excavated or revealed" (Lowe 2005, 20), our interlocutors and informants are prepared to lead us to more open fields. In the next chapters, I consider how these open fields emerge in the branching of networks across scales.

NOTES

1. In Spanish, training is *capacitación*. I attended a number of events like this, and only gradually developed a sense of the subtle differences between *capicitaciónes, educación, fomentación, talleres, seminarios, gestión, fortilización*, and the suite of other terms Chileans use to refer to different types of educational activities. In general, *capacitación* refers to trainings intended to confer a capacity or skill to a subaltern group, especially in the context of empowering *campesinos* to participate in the various bureaucratic forms and processes of development work. To my ear it has a paternalistic tone.
2. ISO 14.000 is a standard for voluntary certification of environmental management systems developed by the International Organization for Standardization, a coalition of national standardization bodies from 162 member states. ISO standards provide voluntary guidelines for national standards governing everything from software protocols to the sizing of industrial components (Sheldon 1997).
3. Mapuche activists often asked me what *kind* of anthropologist I was. For the record, the correct answer for this audience is "political"; "cultural" anthropology they equate to *folklorización*, exoticism or Orientalism.
4. Hartwig has been an important figure in Chilean forestry since the 1970s, and one of the key promoters of the plantation model. His book, *La Tierra Que Recuperamos* (1994), remains one of the key pieces in the argument for plantations as "replacement forests" (see Chapter 1, and Clapp 1998a). His employer, Forestal Mininco, was one of the timber companies most implicated in conflicts with Mapuche communities in the mid-1990s—hence my surprise, imagining Pablo in his classroom.

5. In Stoler and Cooper's terms: "We ask as well how the ways in which colonial states organized knowledge constrains the scholar who returns to those archives (oral as well as written) in an attempt to analyze the colonial situation. . . . We are concerned here not only with the ways—complicated as they are—in which colonial regimes regulated sexuality and biological reproduction but also with how categories of race, class and gender helped to define moral superiority and maintain cultural differences that in turn justified different intensities of violence. . . . With a founding premise that social transformations are a product of both global patterns and local struggles, we treat metropole and colony in a single analytic field" (1997, 4).
6. Lowe's attempts to educate divers about critical injuries or even deaths from the 'bends' were thus met with little response by conservationists—the field of reason that they articulated necessarily eliminated the trace of superstition and framed its practitioners as risks to biodiversity.

4 Green Lungs

> The form of wood, for instance, is altered if a table is made out of it. Nevertheless, the table continues to be wood, an ordinary, sensuous thing. But as soon as it emerges as a commodity, it changes into a thing which transcends sensuousness. It not only stands with its feet on the ground, but in relation to all other commodities, it stands on its head, and evolves out of its wooden brain grotesque ideas. (Marx 1967, 163)

Using firewood to think about society doesn't have a very distinguished past. In our graduate seminar on the history of anthropological theory, we were told of Gerald Berreman dismissing linguistic studies of firewood terms as "a science of the trivial" (Berreman 1966, although the actual source of Berreman's phrase remain elusive). The room full of earnest young anthropologists was aghast. How could one suggest that fuelwoods, on which the vast majority of the world's population depends for all their heating and cooking needs, are a trivial issue? Surely no ethnographer who had ever spent a night in the bush would make such a mistake. To be fair to the venerable anthropologist, he was trivializing the *taxonomic study* of firewood *terms*, a rather different story. So perhaps we were unfair to Berreman, but the firewood's triviality raises real issues about production and consumption, in which the local and domestic is placed in opposition to the global and commercial.

My ethnographic encounter with firewood, and 'certified' firewood in particular, came about as a serendipitous tangent. On the face of it, a local program certifying firewood for local consumption seems like a very different situation than the large-scale, corporate export, global forest issues that I deal with elsewhere. I took up the issue because I was interested to see what it might do to support or trouble some of my ideas about global connections, with different types of consumers, producers, and issues at hand? Since firewood is about heat and cooking, does that aspect of embodiment make for different dynamics than the more abstract nature of mass timber as a commodity? An element of reflexivity is important in this account as well; I became a consumer of local firewood and spent a significant piece of each day worrying about it, carrying it, and trying to burn it.

We all consume. Even that shrinking portion of humanity who consume only what they hunt and gather still participate in, and depend on, the collection and absorption of things. As I argue in Chapter 2, the flows and forms of consuming help shape how we understand ourselves individually and collectively. The small contribution of this short chapter is to imagine out loud that the local and embodied consumption of firewood leads to qualitatively different imaginings of the social relations that enable it—in contrast to the global circuits of pulpwood and building timber that seem so distant from our bodies. The concept troubles some of my assumptions in the other chapters about the formative power of bureaucratic institutions, bringing in a specter of the visceral. Although the bureaucracies of local firewood construct themselves in the shadows of global forms, the discursive politics at work feature unique complexities. The firewood certification project explicitly leverages the public health dimension of 'good' (i.e. clean, dry) firewood to promote 'good' (i.e. sustainable, conservationist) management of native forest; they are quite literally networking lungs to forests. While the firewood project stumbles on some of the same structural economic issues that confront forest certification more generally, the active engagement with firewood's more visceral qualities opens a space for new connections between consumers, producers, and landscape.

Figure 4.1 Lighting a fire (photo courtesy of the author)

I arrived in Valdivia at the beginning of my fieldwork period in the early spring. The south of Chile is cold, gray, and rainy—comparable in many ways to the Pacific Northwest. My wife and young daughter had just endured a week in a small hotel room in Santiago, watching hours of Chilean television while I attended lengthy Fulbright orientation activities. Now we found ourselves in another small hotel room, much colder and wetter, with only the local newspaper to help us find somewhere more hospitable to spend the rest of the year. For three days I beat the bricks in search of housing; none of my university contacts were helpful, and the prospects I saw made me steadily more depressed. Eventually I located the small but tidy and attractive *cabaña* that would be our home.

Signing a lease without any of the proper Chilean identity credentials was a bureaucratic hassle, but as I headed back to the hotel to collect my family, my mind was elsewhere. It was decidedly cold and rainy, and right smack in the middle of our new home was a squat, matte black woodstove. Seeing it, I entertained some nostalgia for childhood winters melting the ice off my mittens in front of my parents' woodstove. But I also began to worry, and I continued to worry all the way back to the hotel. For one thing, how would we keep our child from burning herself? And an even more pressing problem—where would the firewood come from?

Back at our shabby hotel, I consulted the woman sweeping the hallway. I'd seen the grocery store and the hardware store, which catered for most of our necessities, but I had not seen anything resembling firewood. At the same time, smoke was rising from every chimney in town. She seemed a bit put out by my question; this was a modern hotel, with electric baseboard heaters—did she look like the kind of person who knew about firewood? Still, she did have one suggestion, although hardly one that I'd expected.

"Make sure that you buy certified firewood," she told me. This was the first I had heard of certified firewood. I'll ask you, the reader, to imagine my mental condition after three days of apartment-hunting in a strange city in a second language, while my family waited impatiently in a hotel room. I was bemused, or thrown off balance at least, to hear that my best bet for keeping my family warm somehow intersected directly with the research project that brought me here in the first place; and that I had no idea what she was talking about.

"Certified firewood," I asked, "certified for what?"

"I don't know," she replied, sweeping on down the hall, "certified that there's no bacteria in it or something. For the health."

As it turned out, I did not buy certified firewood. While we were moving out of the hotel, I stumbled across a flyer taped to a store window advertising firewood by the sack. I made a phone call and a taciturn man brought out mossy feed sacks full of wood the following afternoon —after we spent a cold first night in the *cabaña*. He had cut the wood on his own property on the outskirts of the city, and brought it loosely piled in old feed sacks in the back of his battered pickup truck. His young son helped us stack the

Figure 4.2 Sacks of firewood (photo courtesy of the author)

feed sacks under the eaves of the *cabaña* before taking my money—about $2 per bag.

My new landlord and next-door neighbor, on seeing my bags of firewood, lamented that I hadn't consulted with him first. He had the inside connection on top-quality firewood, and he was quite sure he'd be ordering some soon. You have to know the right people, he explained, these guys that have to put up flyers and sell by the bag just can't be trusted. A real firewood supplier sells by the *metro ruma* [cubic meter] and has a list of steady clients he delivers to. He also offered me the use of his shed to store the wood out of the incessant rain.

I was not inclined to take my landlord's connections too seriously; it sounded too much like the claims of undergraduate pot aficionados or my father-in-law's boasting about fresh fish. But bright and early the next morning, as I set to the task of starting the first fire with the new batch of wood, I began to see what he was talking about. The wood I had just bought was white pine, not the most efficient wood for heating, and it was green. So green, in fact, that I suspected my vendor had cut the trees down after I called and bagged them immediately. That first fire took nearly two hours to get started, and even then burned fitfully and smoked so aggressively that we had to open windows to chilly damp air outside.

I developed a lot of resentment around those smoky mornings; hunched over in front of the woodstove, poking endlessly at struggling embers, huffing gigantic lungfuls of smoke in my efforts to blow a fire into being. I feel the need to point out that I'm not simply whining about the inconveniences of fieldwork here, but wrestling with a significant health problem. My daughter was just over a year old at the time, and trapped inside in the rainy weather we all quickly developed persistent colds. The damp and the smoke certainly aggravated this condition, but also point to the bigger health issues associated with fuelwoods throughout the developing world. Worldwide, nearly three billion people rely on the indoor use of solid fuels for heating and cooking; the health complications resulting from the associated indoor pollution lead to more than one million deaths annually from acute respiratory infections or chronic pulmonary disease (Ezzati 2005; Ezzati and Kammen 2002). The threat posed by smoky old stoves like mine, to both lungs and forests, has drawn significant international attention; the percentage of a population burning solid fuels is one of the indicators for the Millennium Development Goals (Rehfuess et al. 2006). Replacing inefficient stoves, or replacing biofuels altogether,[1] has been an ongoing project for aid agencies and development programs for decades. Over time the frustrations of transforming this fundamental piece of household technology have provided a sort of case study of the challenges and contradictions of development, 'appropriate technology,' and technology transfer (Barnes et al. 1994; Grace and Arnoux 1998; Mehta and Shahpar 2004; Subramaniam 2000). Although Chile is a relatively affluent country by the standards of the Millennium Development Goals, respiratory infections attributable to indoor pollution are a serious problem for poor Chileans in urban and rural areas alike (Adonis and Gil 2001; OECD 2005). Little did I know that I had introduced my family to one of the world's great public health problems. The visceral experience of fieldwork, that sense of physically 'being there' that makes ethnography what it is, intersected unfortunately with a visceral reality that affects millions.

Fortunately, I eventually worked out an ad-hoc solution. Using vast quantities of newspaper and other scraps, I would start hot, fast fires—as I commented in Chapter 1, this was a stage of the process that generally required a lot of wrestling with my eager toddler. Meanwhile, I would stack the green logs directly on top of the woodstove to dry out before being jammed in the stove. This strategy had its risks; if not watched closely and moved around regularly, the drying logs would begin to smoke and even eventually smolder into flame. Still, I had found a way to keep burning the sub-par wood to keep the house warm without completely asphyxiating us all. Once 'kiln-dried' in this fashion, my wood would quickly light and burn with little or no smoke.

Before the cold, damp early spring had completely given way to summer, the landlord's firewood connections finally resurfaced. He cut me in on a large truckload of wood with such theatrical duplicity I honestly believed he wanted me to smuggle Cuban cigars into the States or something equally shady, but we eventually reached an understanding. Some

Figure 4.3 Kiln-drying the firewood (photo courtesy of the author)

days later, without warning, a flatbed truck rumbled up onto the lawn. One man began heaving meter-long logs off the back of the truck, while another stacked them in a row on top of the flower garden. Once a rough stack had been formed, both men set to work with chainsaws, scrambling perilously across the now-sprawling pile of wood. Meter-long logs were roughly split lengthwise, then cut in half to make half-meter stovelengths. The pile represented a variety of species, by my estimate—white pine, but also "*pino oregón*" [Douglas fir, *Pseudotsuga* spp.], *ulmo* [*Eucryphia cordifolia*], *roble* [beech, *Nothofagus obliqua*], and even sweet-smelling apple, among others I couldn't readily identify. Even before the splitting, I could see that the logs had clearly been cut last year and left under cover to dry.

As the cutting progressed, a set of social networks materialized that I couldn't begin to trace. One of the men with chainsaws was the landlord's cousin; the land where the wood had been stored belonged to another relative, while the property where the wood was cut remained a secret participants would joke about but not reveal. The ownership of the wood was as partial as its provenance. I'm not sure how many of us had bought shares in the truckload, but people came and went all day with smaller trucks and wheelbarrows to portion out wood; the sub-pile in this photograph actually sprouted at a neighbor's house.

Although there was a great deal of talk about money, mine was the only money that I saw change hands. When I expressed amusement at the goings-on, participants told me that this sort of chainsaw party was the normal method for getting firewood in urban areas throughout Chile. As spring turned to summer, I often saw similar undertakings in other neighborhoods.

This wood was not certified. Based on the general elusiveness about where it came from, I'm not entirely sure it was legal. Likewise, the green wood I bought by the sack was not certified, nor was the somewhat better wood I later bought by the sack from the grocery store. Certified firewood, it turned out, did not in fact *exist* as yet anywhere in Valdivia. But it was on its way, and signifiers of its presence were everywhere. As my research progressed and I became personally acquainted with many of the professional

Figure 4.4 Firewood delivered and divided (photo courtesy of the author)

foresters in the area, I found myself discussing the new Sistema Nacional de Certificación de Leña with nearly everyone I interviewed. I even learned that firewood has a website in Chile: www.lena.cl. It was from the website that I learned about the event launching a version of the certification system in Temuco that I describe below. But the most visible signifier of certified firewood was Lenny the Leñito, seen here on the side of a *cabaña mobil* set up in Valdivia's central plaza on a sunny summer afternoon.

I still regret deeply that the cabin was closed up the only day I had my camera handy: On the inside is a carefully designed display of different kinds of qualities of firewood, with instructions for how to properly start a fire and a working woodstove for the staffer to demonstrate the process. The *cabaña* was staffed by a junior member of the Agrupación de Ingenieros Forestales por el Bosque Nativo [Organization of Forestry Engineers for Native Forest], one of the NGOs at the vanguard of Chilean certification—both that of firewood and of timber more generally. FSC-Chile President Luis Astorga was a founding member of the Agrupación. While FSC certification and the Sistema are completely different projects, they share a great overlap in personnel.

The Consejo de Certificación de Leña (the coalition of NGOs and government bodies promoting the Sistema de Certificación de Leña) or COCEL handed out fliers and posters in Valdivia and Temuco informing the public that

Figure 4.5 Cabaña mobil (photo courtesy of the author)

"*La leña en Valdivia/Temuco pronto tendrá un sello que será garantía de:*

- *Más beneficícios para los consumidores*
- *Leña seca*
- *Comercio legal y transparente*
- *Menos contaminación*
- *Uso responsable del Bosque Nativo*"

That is, "Firewood in Valdivia/Temuco will soon have a seal that will be a guarantee of: more benefits for consumers, dry firewood, legal and transparent commerce, less contamination, and responsible use of native forest."[2] These claims represent the stipulations required of vendors participating in the *leña* project, codified into very particular terms. "More benefits for consumers" refers to the requirement that vendors sell wood in units of a standard size, accompanied by a bill of sale indicating species and humidity content of the wood. "Dry wood" means a humidity content below 25%, as measured by an electronic meter. "Legal and transparent commerce" refers primarily to the *guía de tránsito* [transit invoice] stamped by CONAF, which documents the source and proper custody of the wood. Such an invoice is already required by law; ostensibly a truck full of firewood entering the city could be stopped by police and asked to show their invoice. As fond as Chileans are of their paperwork, none of the firewood vendors I encountered in my less-than-systematic survey that summer had ever secured a proper invoice.

Most interesting to me are the claims for "less contamination" and "responsible use of the native forest." Each of these represents an attempt to link the concerns of firewood consumers to what one of the Agrupación staffers told me were the "*asuntos mas grandes, mas importantes al nivel mundial,*" [the bigger issues, more important at the global level]. "Contamination" as a concept encompasses "indoor pollution," which as I have mentioned is a severe threat to public health—and one with which householders in the south of Chile are extremely and intimately familiar. It also includes outdoor pollution, the quality of air at the municipal or even regional level. Southern Chile has a cool and damp climate, at least on the coast; widespread use of wood for cooking and heat puts huge amounts of particulate matter into that air. There, it tends to linger and mix with automobile exhaust to create smog. "Less contamination" then bridges the space between the household consumer of firewood and the community as a whole via a discourse of public health. The same Agrupación staffer was explicit about the intent of the *leña* outreach project to *concientizar* [raise the consciousness of] the public about urban air pollution by making the connection to household air quality. "Responsible use of the native forest" involves another *concientización* project. According to my informant, the general public in Valdivia was not deeply concerned about the state of the native forests—they didn't represent a salient issue for most urban dwellers.

By associating forest conservation with recognized values like health and fair business practices, the Agrupación hoped to gain a small regulatory foothold while increasing public awareness.

I spent several weeks of my time in Valdivia pestering the president of the Agrupación de Ingenieros Forestales por el Bosque Nativo. At 28 years old, Rene Reyes was extremely young to hold such a leadership role, yet he'd already been actively engaged in a huge range of sustainable forestry projects throughout Chile (and elsewhere in South America) for years. As I became more curious about the *leña* project, I was eager to speak with him about it, but also because of his role in Chilean forest politics more generally. Reyes had been a friendly critic of FSC-Chile from its outset, and in conversations at public meetings had seemed doubtful about the entire principle of certification as a method for promoting sustainable forestry via that market. So I was interested to hear what the *leña* project meant to him—unfortunately, it was a conversation that would never take place. I had told Reyes at our first meeting that I was in Chile to study the FSC; since he had no official role in FSC-Chile, he simply couldn't fit me into his schedule.

So I felt fortunate and surprised that, when COCEL announced a press conference and reception to celebrate the launch of the Sistema in Temuco, Reyes invited me to join him. We drove to Temuco in a small car packed with six *ingenieros* and myself. The launch event took place in Temuco's six-story city hall, in a press room with TV cameras and a catered snack table. A parade of dignitaries proceeded across the podium, prepared speeches were read dryly, there were slideshow presentations with graphs of particulate matter in Temuco attributable to woodsmoke, and so on. A team of development staff from the German aid agency, GTZ, were on hand, describing the role of German aid money in the appraisal and implementation research, and more funds later to be channeled through Fundación Chile (the country's largest funder of non-profit initiatives) and municipal governments. Several players made the argument that bringing the fate of native forests into the homes of urban consumers will raise conservation awareness, while linking public health to forest conservation in the minds of funders and the general public.

At the time, struggling to maintain focus in a hot room full of mumbling bureaucrats, I was struck by the contrast between this event and the chainsaw party in my yard. The success of the firewood certification program overall, though, depended on the ability to bridge the gap between bureaucratic standards and convivial resource-sharing. COCEL's plan to raise the green consciousness of firewood consumers by linking healthy forests to healthy lungs partakes of the logic that Agrawal (2005) calls "environmentality": That environmental regulation succeeds through interpellation, through leading publics to recognize themselves in a management discourse.

What even the casual observation of local firewood certification throws into high relief is the awkward disconnect between standards and practices. The point is not so much that timber producers and sellers are not practicing responsible forestry in the fashion prescribed by standards. What struck me was how the technocratic imaginary of the organizations and officials promoting the project sits at such a great remove from the understandings of the producers, vendors, and consumers involved. The paperwork required to legitimize a certified firewood vendor seems to work at odds with the kind of dense social networking I encountered at the chainsaw party on my front lawn.

It also directly rules out the kind of enterprise that brought me the feedsacks of green wood on my first day in the new *cabaña*. This is a good thing, on the whole; burning that kind of crappy green wood contributes to pollution in the skies and in homes, with deadly effects. What troubles me in the *leña* certification project, and its associated outreach, is the implicit assumption that consumers need to be told if wood is dry enough to burn, and told that burning green wood is bad for themselves and others. If I can make any generalizations from my limited experience crouched in front of my stove, there is no one that needs to be told that green wood is bad. Low-income urban and rural Chileans burn green wood for a number of reasons (green wood can burn *longer*, for example, thus keeping a fire smoldering overnight), but I cannot imagine that ignorance of its qualities is among them. Cost is undoubtedly a factor, especially in terms of time and ability to pay; poverty deprives many consumers of the efficiencies of scale. That is, I know that the elderly woman who lived down the street from us in Valdivia bought wood by the sack, although this wood is greener and more expensive than that bought in the big chainsaw party. She did not have the quantity of ready cash on hand that she would have needed to buy a share in the large load, and instead had to buy what wood was available in small sacks when her funds allowed. Since meeting standards costs money, certified firewood is inevitably more expensive than un-certified (Haener and Luckert 1998; Sedjo and Swallow 2002). This does few favors for the already marginal who face the smoking woodstoves every day.

I conclude on a comparative note: What does the experience of the firewood certification system signify for a global project like that of FSC-Chile? To begin with, both programs face a challenge of external validity. In the case of certified firewood, is it really addressing the problem it sets out to solve? Making the link between urban consumption and the state of Chilean native forests is one project, one that seems well-articulated in the certification scheme, but preventing indoor and outdoor air pollution is a very different issue. I wonder here, as I wondered in conversation with foresters in southern Chile, if certifying firewood perhaps drew attention away from other strategies for preventing indoor and outdoor pollution: Heating subsidies, better stoves, enforcement of existing regulations, promotion of alternative heat sources, and so on. Perhaps creating a separate high-priced market for 'good' firewood will harm the existing market for 'bad' wood, driving prices up and producers out of business, lowering quality further, and harming precisely the marginal wood-burners who risk their health the most. On the other hand, some have argued that in inelastic markets like those for most wood products, the costs involved with certification are typically internalized by the producers rather than passed on to consumers (Sedjo and Swallow 2002). In either case, the connection between the economic dynamics set in motion by certification and the outcomes in terms of air, forests, and lungs remains less than clear. FSC-Chile faces a similar conundrum. For all its attention to internal issues of adequately defining sustainable forestry and representing stakeholder interests, its potential impacts on the market are difficult to assess. Certification as a 'non-state market-driven' form of environmental governance is inevitably subject to certain internal tensions—Marx would say 'contradictions.' Creating a value in the marketplace for non-market values results in uncertain, possibly perverse, outcomes. This structural paradox is shared at different scales by the FSC and by the firewood project, and by perhaps by other ethical market initiatives like organic food or Fair Trade coffee (Guthman 2004; Taylor 2005).

In more visceral terms, I am fascinated by the bodily experience of the firewood and what it means for certification. As I noted in Chapter 2, timber and wood products generally don't make very good luxury goods—for the most part we do not experience them with our bodies. As a result, the premiums associated with certifying those products depend on a much more intellectual or even institutional process than, say, organic food. Poor quality firewood, on the other hand, has a very clear and intrusive quality of embodiment, as my rasping lungs would attest. Dry dense wood is a joy to heat with—it lights quickly and burns hot, with little to no smoke emitted into the living space, and hence the lungs. The advertising and extension brochures emphasizing quality, as opposed to sustainability, had material to work with that was logically coherent and easily recognizable to its audience. If the link between indoor pollution and native forests appeared to local consumers as an abstraction, the value of good firewood was visceral and self-evident. Chile's certified firewood project succeeds where the FSC seems likely to fail, insofar as the visceral quality of firewood consumption makes the ethical content of certification real for consumers in a material and embodied fashion.

Haraway has a term for the embodied quality of the ethical quandaries of consumption: She calls it indigestion (Haraway 2008). It is a phrase that resonates in many directions. In some sense we all consume; to live is to eat, after all. In Haraway's treatment, eating is the fundamental act of creating ourselves in a universe full of Others, and it encompasses every aspect of how we carve out space for ourselves. Most tellingly, eating is always inevitably inseparable from killing—even the most ascetic among us cannot make ourselves without becoming complicit in the unmaking of others: "outside of Eden, eating also means killing, directly or indirectly, and killing well is an obligation akin to eating well. This applies to a vegan as much as to a human carnivore. The devil is, as usual, in the details" (2008, 296).

Tracking those details, negotiating a responsible engagement with the others that one lives with and eats (with), can only be a process, not a given. Haraway provides a critique that applies equally to my concerns about certification systems: "There is no rational or natural dividing line that will settle the life-and-death relations between human and non-human; such lines are alibis if they are imagined to settle the matter 'technically'" (2008, 297).

It's that word 'technically' that connects most directly to my thesis about the Forest Stewardship Council, my identification of certification as a technology of value. In place of a technical adjudication, Haraway recommends indigestion (or "nourishing indigestion"), that pointed and uncomfortable awareness of one's consuming relationship with other beings. Ethics are not easy, and articulating the knowledges and scales of forest certification may lead to headaches as well as stomachaches. But I share Haraway's concern about a technical system designed to sell a commodified green innocence to an individualized, depoliticized ethical consumer. In the context of the Slow Food movement, Hayes-Conroy and Hayes-Conroy (2008) argue that a visceral or even aesthetic engagement with knowledge is not merely epiphenomenal, but rather an important building block for political subjectivity and mobilization (see also Hayes-Conroy 2009; Gabrielson and Parady 2010). If environmental knowledge is best understood not as a cognized lexicon but as a phenomenological event (as I discuss in Chapter 3), then this model of embodied politics makes all the more sense. Chile's firewood project seems to suggest that visceral experience—the smoky lung—may serve as a point of connection between constituencies and fields of action—one that puts physical weight behind the abstractions of standards and certification.

ACKNOWLEDGMENTS

A version of this chapter was originally published as Henne A., 2010, "Green lungs: good firewood, healthy air, and embodied forest politics." *Environment and Planning A* 42(9) 2078–2092. Pion Ltd., London. www.pion.co.uk and www.envplan.co.

NOTES

1. That is, with kerosene or propane, or power from the electrical grid, any of which have ecological and social implications far beyond the scope of this project.
2. Note, though, that despite the chambermaid's comments that introduced me to certified firewood, bacteria content is nowhere on the agenda.

5 Certification and the Politics of Scale

As an ethnographer, my preferred data-gathering method is the "semi-structured" interview. That is, I have not used a set of survey questions or a standardized interview protocol in any of the interviews that went into making this book. In general I've been happy to let the people I speak with direct the conversation according to how they understand my questions. That said, I do come into the conversation with a certain set of goals, things I want to figure out; how does a certification audit work, what was your role in negotiating the current standards, did you go to this meeting? For the most part, this strategy worked well for me, and I would walk away from a conversation with a tape-recording in which my questions were more-or-less answered. Still, there were many occasions when I found myself struggling with a certain kind of misunderstanding about scale. The example that comes most readily to mind took place during an interview with a board member of FSC-Chile about a protected area project his own organization had recently proposed. The proposal was somewhat vague, but it included some of the communities that were also planning to participate in FSC certification, so I asked how it had come about. I think my exact words were, "¿*Cómo empezó este parque?* [How did this park begin?]" My interlocutor cleared his throat, took a drink of water, and began to lecture. What one needed to consider, he told me, was the pre-Colombian nature of peasant agriculture in Chiloé, from which point he elaborated on the historical and political agroecology of the region since the Conquest, for better than 15 minutes. When he finally paused for another drink of water, I was able to clarify, "Right, but I meant, who called the first meeting, you or Don Pedro?"

While these miscommunications may stem from habits of pontificating common among scientists or academic types everywhere, I found that getting answers to specific questions required real persistent effort; names, dates, and facts were rare, while extensive explanations of context or meaning were the norm. As I gained confidence I became more comfortable interrupting, and I eventually learned to navigate the issue pretty well in interviews and conversations. But the fact that I had to work so hard to rein in the abstractions in my interviews suggests to me that scale is a critical issue in the design of standards and in forest certification as a practice. For my interview subjects, it was genuinely difficult to answer questions about who-did-what-when-and-why without resorting to explanations across large scales of space and time. As a regulatory form that explicitly links Northern consumers with Southern producers, forest certification is heir to a whole set of scalar complications. It is not unreasonable to suggest that the thorniest aspect of environmental regulation is the selection of the appropriate scale for intervention (i.e. Reid et al. 2006). For this reason, I begin my discussion of scale in Chilean forest certification with a brief review of the literature on the politics of scale in geography and political ecology.

It would probably be too strong to say that political ecology is isomorphic to the politics of scale. The concerns of the latter undoubtedly form the backbone of the former, though, and I do think it offers at least a good point of departure for discussion. One of the foundational texts in political ecology is Blaikie and Brookfield's (1987) *Land Degradation and Society*, which offers the strongest theoretical statement of arguments found elsewhere at the time including Schmink and Wood (1987) and Sheridan (1988). According to their classic definition, political ecology consists of a combination of "the concerns of ecology and a broadly defined political economy. Together this encompasses the constantly shifting dialectic between society and land-based resources, and also within classes and groups within society itself" (Blaikie and Brookfield 1987, 17). The works of political economy that are most often referred to in this classic formation of political ecology (besides Marx, of course) are quite spatial in their perspective, whether they emphasize world economies (Cardoso 1972; Frank 1975; Wallerstein 1974) or imperialism and culture (Mintz 1985; Wolf 1982). In all cases, the decisive location for social and environmental processes is the intersection of the local and global scales.

There are two dynamics that Blaikie and Brookfield describe as key to environmental degradation, both quite spatial and therefore informed by scale. *Marginality* refers to the tendency for politically powerless people to be restricted to poorer lands, forcing them into economically tenuous positions that lead to ecologically harmful behaviors. The driver for this process, typically market forces or state intervention, can be traced along *chains of causation* from the local actor to the highest level of the global market or vice versa. Vayda and Walters (1999) are troubled by the *a priori* quality of the chains-of-causation argument, and propose in its place a *progressive contextualization* of environmental events; a particular driver is not necessarily located at either end of the scalar chain, but the analyst reframes the event in its relevant contexts at increasing larger scales. In both cases, the linear path of analysis from the local to the global is assumed to follow a real-existing series of nested scales—this is where the geographical literature on the politics of scale can make a much-needed intervention.

'The local' is a vexed construct in political ecology, and has been from its beginnings. Blaikie and Brookfield's text is built on analyzing the behavioral patterns of a hypothetical 'land manager'; a peasant farmer or herder or woodcutter who represents the iconic local actor. The key goal of Blaikie and Brookfield's project is defending this local actor against accusations of an environmentally degrading irrationality—they demonstrate that, given the land manager's situation in the midst of shifting markets, falling prices, insecure tenure, and an interventionist state, environmentally degrading practices (overgrazing, shorter fallows) are perhaps the land managers' *only* rational choice. Explaining environmental degradation takes the form of a chain of causation *away* from the local land manager into successive levels of complexity at larger scales—from *proximate* to more *ultimate* causes.

The perspective is revolutionary, when considered in the context of Malthusian environmental models that blame increasing populations and the voracious irrationality of the poor for the loss or degradation of forests, soils, and water. With the benefit of hindsight, though, the shortcomings of this linear chain of explanation are clear. In clearing the local actor, the land manager, of blame in environmental degradation, the classic model of political ecology also underrates his or her capacity for agency. Are all actors really so tightly bound by social and political forces acting at larger scales? Vayda and Walters (1999) question the scientific validity of the premise, arguing that some cases of environmental degradation might really be the fault of irrational management or other local variables, and that an *a priori* chain of explanation puts the cart before the epistemological horse. I think there's a certain merit to Vayda and Walters's argument, but I prefer to deal with the complications of local agency by questioning the categories that Blaikie and Brookfield, and many others, rely on for these explanations. Fortunately, there are plenty of political ecologists covering this ground already.

To begin with, the 'land manager' is something of a theoretical fiction derived from classical economic models—the rational individual making self-interested choices, optimizing efficiency by maximizing benefit while minimizing costs (see Boyce 1994 for some discussion). In economic anthropology, 'the household' is often used as the basic unit of analysis inserted into these models, since property holding and consumption patterns tend to converge at that level. Interesting work in political ecology has taken place around what Robbins (2004) describes as "breaking open the household." Dianne Rocheleau (1995; Rocheleau and Edmunds 1997; Rocheleau et al. 1996), Richard Schroeder (1999), Jill Belsky (2003), and Judith Carney (1996), among others, have delved into the role that gender divisions play in resource management at the household and local community levels. It turns out the rational interests of the land manager might vary quite dramatically depending on his or her gender. In Schroeder's definitive *Shady Practices*, he documents conflicts arising between men and women within communities and households over access to valuable horticultural land, as development interventions interfere with traditional management practices. Men were encouraged by development institutions, and by high prices, to plant fruit trees wherever light and irrigation were best—which happened to be in or near women's market gardens. Carney demonstrates a similar pattern in terms of wetland rice cultivation and men's and women's different participation in market agriculture; Belsky discusses the part that the gendered division of labor plays in the conflicts arising from an ecotourist project.

Another good example comes in the work of Fairhead and Leach (1995) on forest island management in Kissidougou. First, they note that the trees on the landscape are not the *remnant* of forests destroyed by local villagers, but the *product* of their active management. Programs geared toward improving or replacing local management are aiming at a moving target, since management of village resources actually takes place across several different scales for different purposes; trees are planted by individual farmers, but trade is regulated by village elders, and grazing rights are negotiated with neighboring pastoralists by an intervillage authority. Identifying 'the local' here is problematic. Donald Moore (1996) offers another take on local politics, or 'micropolitics' as he calls it, in the relationship of Zimbabwean village elders to village farmers and national park authorities. He argues that it would be impossible to understand these actors without knowing the discursive politics emerging from a long and complicated colonial history. In a conference entitled *Representing Communities* (Brosius, Zerner and Tsing 2005), anthropologists and other conservation researchers complicate the notion of the local and the community in regards to community-based conservation measures. As conservationists and resource managers have learned to their chagrin, the characteristics of a community are far from self-evident. "Who's local here?" (pace Peters 1996) is a question recognized as deeply political and discursively complex by political ecologists.

Locality is not the only scalar property subject to problematization. Herod and Wright (2002) argue quite cogently that these locations we know as scales, the local, the regional, the national, the global—are not self-evident places or stages for action. The keystone argument warrants quoting at length:

> Why, for instance, was the September 11, 2001 bombing of the World Trade Center regarded as an act of 'global' terrorism instead of regarded as an act perpetrated within the sovereign jurisdiction of the United States and thus a 'national' or even 'local' (New York City) crime? Was it because pictures of the burning twin towers were instantly flashed around the world? If so, then why are the pictures of war victims in, say, Kashmir or Central America which find their way to the world's television sets not considered to be representations of 'global' events in the same way? Was it because the WTC victims included people from very many different countries of the world? If so, then why are not the kidnappings and murders of foreign tourists and business people in places such as the Philippines, Colombia, Egypt, and Russia not 'global' events in the same way? Was it because of the magnitude of deaths? If so, then

why are the thousands of people who are murdered every year in the United States (an estimated 15,517 during 2000, according the Federal Bureau of Investigation), not to mention the hundreds of thousands in other countries across the planet, not considered likewise to be part of a 'global' news story? How, in other words, do we determine that terrorist cells—or environmental problems or worker abuses or anything else—around the world represent a global threat instead of different kinds of local ones? (Herod and Wright 2002, 2)

This echoes an earlier question provocatively posed by political ecologists, "How do we know we have global environmental problems?" (Taylor and Buttel 1992). To answer the rhetorical question, an issue becomes global (or local, regional, or national) by virtue of specific discursive strategies on the part of actors involved.

In a study of anti-toxics activism in Louisiana, Kurtz (2002) draws on social movement theory to define these discursive processes in terms of frames, specifically *scalar* frames. In her example, activists cast their protest against a chemical plant in terms of increasingly larger scales, in order to incorporate or attract support from specific actors. In turn, supporters of the plant attempted to frame the protesters at smaller scales, to deprecate their concerns or isolate them from larger communities of support. Both tactics were discursive and active projects. Swyngedouw describes the deployment of these scalar strategies as the "politics of rescaling," and argues that "the production of space through the perpetual reworking of the geographies of capital circulation and accumulation junks existing spatial configurations and scales of governance, and produces new ones in the process" (2000, 68).

My experience in and amongst the Forest Stewardship Council in Chile suggests that the project of certification represents a self-conscious entry into the politics of rescaling on the part of many of its constituent actors. My first interview with FSC-Chile president Luis Astorga touched on this issue by way of one of those Big Picture narratives I mentioned earlier. I began by asking, "*¿Al princípio, cómo llegó certificación a Chile?*" [How did certification first come to Chile?]. He responded with a 15-minute lecture, in this case neatly explaining the marginalization thesis (see previous) and its relation to colonialism, modernization, and development from 1492 through the military dictatorship. He wrapped up, though, by adding that the FSC intended to reverse the process, by "*tomando valor del mercado internacional y agregandolo en lo marginal*" [taking value from the international market and adding it to the marginal]. Astorga's comment locates certification as a self-conscious scalar project, attempting to cast the particular scales as ontological fixed reality while eliding the very particular practices and network effects that make scale what it is within FSC. The 500-year scope of his political ecology also reiterates the issue of temporal scale. Below I elaborate on some of the scalar practices and implications of certification in Chile.

MAKING THE SMALL PROPRIETOR

The final version of the Chilean national standards that FSC-Chile sent to FSC-International for ratification is laid out in a tabular format. The rows count down through the 10 Principles and Criteria, broken up into relevant indicators and verifiers. The entire table is divided into two columns: "large scale" and "small scale." Each Principle, each Criteria, each indicator and verifier is defined in terms of its relevance to forest operations at large and small scales. In many cases the language in each field is identical, i.e. Principal 3, Criteria 1, indicator 3: "*No se manejan las areas de conflicto establecidas en mapas, al menos que exista una clara evidencia de un libre e informado consentimiento por los indígenas y la población tradicional del área,*" [Areas where conflict is established by maps will not be managed, except where there exists clear evidence of free and informed consent by the indigenous and traditional population of the area]. Often the columns appear similar but clearly have been altered to reflect the different priorities and capacities of larger and smaller operators, i.e. Principal 1, Criteria 3, indicator 2 reads, "*Existe una estrategia implementada en el PMF dirigida a generar condiciones de estabilidad laboral a los trabajadores,*" [An implemented strategy exists in the project directed at generating conditions of stable employment for the workers] appears in the 'large scale' column, while the 'small scale' column reads, "*No existen evidencias de discriminación, en todas sus formas, y en especial en relación a la igualdad de oportunidades y trato en el empleo de acuerdo de al Convenio 111 de la OIT,*" [There is no evidence of discrimination, in all its forms, especially in relation to equal opportunity and treatment on the job in accordance with ISO convention 111]. In other cases, text appearing in one column is entirely absent from the other, i.e. Principal 4, Criteria 1, indicator 1 appears only in the 'large scale' column: "*Existe un compromiso y esfuerzos documentados para favorecer la contratación de mano de obra local y si es necesario se dará entrenamiento a la comunidad para incentivar su contratación,*" [Commitment and documented efforts exist to prioritize hiring local labor, and if necessary training will be given to the community to 'incentivize' their hiring].

This was not the case all along. FSC-International's Principles and Criteria are not divided in this fashion, nor was the original draft Chilean standard prepared by Luis Otero (see the following). The division of FSC-Chile's standards into 'large scale' and 'small scale' reflects a set of concerns and identities that are not uncommon in forest politics worldwide, but that have particular resonance in Chile. I refer in particular to a subject position that appears in a variety of important discourses in Chile, the *pequeño propietario forestal* or "small forest proprietor." Not unlike the "family farmer" of US political discourse in the 1980s, the small proprietor sometimes seems more like a state of mind than a demographic fact. Here's the definition of the small forest proprietor from the Chilean legal code (Decreto Ley 701 de 1974):

> The person who, in addition to the requisites of the small farm proprietor defined in article 13 of lay 18.910, works and is proprietor of one or more rustic properties, whose combined area does not exceed 12 hectares of basic coverage . . . In all cases, those properties will be considered not to exceed the equivalent of 12 hectares of basic coverage that have an area of less than 200 hectares when located in Regions I through IV, XI, XII, in the community of Lonquimay in the IX Region and in the province of Palena in the X Region. Included among the small forest proprietors are the farm communities regulated by legal decree Number 5 of 1968, of the Ministry of Agriculture; the indigenous communities covered under Law 19253; dry-land societies constituted according to Article 1 of Decree Law 2.247 of 1978; and the societies referred to in Article 6 of Law 19.118, assuming that at least 60% of the collective capital of said societies remains in the hands of the original members or of persons who have the quality of small forest proprietors, accordingly certified by the Farm and Livestock Service.

If the bafflingly precise legal definition appeared in 1974, the concept has been around quite a bit longer. Camus (2005) locates the birth of the Chilean "small forest proprietor" with the arrival in Chile of Germanic 'scientific forestry' in the early 20th century. German foresters traveled to Chile, as they did to many timber-producing countries (Peluso 1994; Vandergeest and Peluso 2006; Wiersum 1995), at the request of timber companies and the Progressive government; they offered technical support to the expanding timber industry and development recommendations to the state. The small proprietor emerged as a target for interventions aimed at scaling-up, assimilating smaller properties and operations into larger ones managed by cooperatives or privately owned companies. Towards the end of the 20th century, depictions of the small proprietor shifted away from their inefficiencies to emphasize the value of their native forest patrimonies. One of Chile's more famous conservationists mentioned to me several times the importance of conservation outreach to this population, because 70% of Chile's surviving native forest was in the hands of small proprietors. The best evidence I can find suggests that percentage is closer to 10% (Centro de Análisis de Políticas Públicas 2006). The difference between the documented and imagined figures may have something to do with problematic definitions of what constitutes native forest (Neira et al. 2002), or with vague measurements of the number of small proprietors themselves (Otero 1999). In either case it reflects the significance that many actors in the forest sector place on the iconic small forester and their valuable/vulnerable patrimony. The pressure to consolidate small properties has been replaced by a variety of sustainable development extension activities and *capacitaciones* intended to help small proprietors conserve native forest by marketing its products (Bluth Solari 2002; Catalán et al. 2006; Camus 2005; Casals 1999).

A recent example of the genre was produced by *ingeniero forestal* Luis Otero, a founding board member of FSC-Chile. His *Manual de Manejo Sustentable de Bosques Nativos para Pequeños Propietarios [Manual of Sustainable Native Forest Management for Small Proprietors]* (Otero 1999) was published by CODEFF, the lead environmental NGO in FSC-Chile. Lovingly illustrated with the author's own watercolors of forest landscapes and foresters at work, the manual provides both practical advice on managing native species for market and a theoretical introduction to forest succession and disturbance regimes written for a non-expert audience. What makes the book fascinating is the rhetorical device by which Otero locates his advice for small proprietors: A fictional family of small forest proprietors named *la familia* Nahuelpan Gonzalez. They have conversations like this:

> "We have a few hectares of cleared land that we could use for plantations," said Juan, "that's good business, and it's not expensive; we could sell a couple of lambs, plant and then the Revenue gives us back the money in the forest subsidy." Filomena, who had been through the fields looking for mushrooms and fruits, knew that it was full of second growth, and that it wasn't worth the trouble to sell livestock to plant pine when there was so much beech. "There's beech growing all over, why would we want to plant pine? If we take care of the second growth we'll have the best wood. A few good pieces of laurel, beech or hazel would fetch a good price, wouldn't they?" Grandmother Flor agreed. She has seen the forest burned and sprout back, and for her the forest represents part of her culture, especially the laurel, sacred tree of the Huilliches. For Enrique, who would finish school soon, the woods were a place for games and soon would be a place for work. He is interested in the issue of the environment, since his teachers at school talk about this a great deal.

All the elements of the small proprietor identity (including indigeneity) are present and cast in relation to their role in making forest-management decisions. The depiction is probably a realistic one; certainly Otero has the experience with the population in question to support his illustration. What's compelling about the manual isn't the quality of the representation, but rather it's application in a document intended in the end to endorse a particular land-use; management of native forest rather than conversion to plantation. Even if the manual's language is carefully non-expert, it's clear that its intended readership is not the small proprietors it depicts[1] but the technical professionals working with them. Effectively, the manual's message is not "how to manage native forest," but "how to train and support small proprietors in managing native forest."

Throughout, the small proprietor is imagined in relation to external forces, either those attempting to assimilate him out of existence or to enroll him in the conservationist project. I spent a certain amount of time participating

Certification and the Politics of Scale   103

in this process myself: In my capacity as a volunteer with CODEFF I helped prepare a manual of forest laws for small proprietors, distilling Chile's dense pile of forest laws into a readable list of relevant codes. I don't have much to say based on my fieldwork about the lived experience of the small proprietor (although see Catalan et al. 2005), or whether there is some authentic quality behind the representations deployed strategically by other actors. I would like to note, though, that my conversations with small proprietors revealed a reflexivity to the relationship; that is, they too construct themselves as small proprietors in relation to larger economic forces. Here's what "Maria," one of the foresters in the association of small proprietors with whom CODEFF worked, had to say when I asked her if she preferred the life of the 'small proprietor':

> Yeah [laughing], yeah. I don't, uh, while I have the strength and the health I'm going to give the best example I can to my kids so they can follow the same steps. Up 'til now at least they're good . . . too soon to tell, right? I can't say, I can't give them presents and let them do what they want. No, I'm sort of old-fashioned. But I tell them, "kid, we have to, you just have to do it, because that's how it is." It's like when they gave me the sustainable management plan, damn . . . I wish I'd had, I don't know, they'd done something . . . A little more support for the small forester. Because I myself would work more in the forest, or my husband would. In the end he could work more in the woods, but without any income, we can't. Of course we can't, we have to go out to work over there and maybe when we get back we can chase the livestock out of the woods or something, because you know we still don't have a fence. I can't, I don't have the money to keep buying these things, to keep looking after the forest. If I did, I don't know . . . Just a little subsidy, or I don't know, a stipend, say a little something like that. Damn, if they made a stipend for the small forest proprietor like that it would be . . . it would help! It would be, give us that, right, a couple of months and I'd say, "Right! I'll do two or three months of pure forest work and nothing else." I think there'd be a lot more care for the forest, it wouldn't be abandoned like this. And then the other months you could work in town or be free for other things. Just yesterday we finished planting, filling in the forest, and dammit, today I see I'm out of detergent, I don't have salt or sugar—it's depressing sometimes. You might say, "Damn, what am I earning here in the forest; I'm not earning anything and now I don't have things for the kitchen, or bread for my kids either."

On the one hand, Maria is describing the kind of economic crisis common to peasant agriculture around the world, the management of a smallholding with its attendant costs and opportunities, the balancing of productive profits in one sector (i.e. wage labor in town) against the labor costs of entering another (i.e. planting trees for future harvest as firewood). At the same time, she refers explicitly to a 'them,' an outside agency that even in this brief tirade includes her municipal authorities, the state (via CONAF), and the NGOs promoting FSC certification among small proprietors. Her identity as a small proprietor, to the extent that it's relevant to her self-concept at all, is defined in relation to these larger entities, these potential sources of monetary support. FSC certification assumes a role in redefining those relationships. Participating in certification encourages Maria to imagine certain NGOs as conduits to larger circuits of value, while lending stability to those aspects of the 'small proprietor' role that mesh with certification's reach—the income, the care for the forest. The scalar aspects of certification as a technology of value come into play here—if Maria imagines the FSC as a piece of her domestic economy, certification becomes a node in a network connecting her to the larger systems, to the international market. Certification derives its legitimacy from interpellations like this one, each new connection in the network smoothing its function and extending its reach.

CERTIFICATION IN ACTION: HOW TO FOLLOW GOOD WOOD THROUGH SOCIETY[2]

Conservationists, timber companies, academics—in our lives we all deal with a great deal of paperwork. Especially in the moment that it confronts us, it's tempting to imagine paperwork as an obstacle to movement, a constraint on our capacities for action. An actor-based approach to environmental politics (i.e. Paulson, Gezon, and Watts 2005) might frame the paperwork of environmental regulation as an epiphenomenon of the regulatory process. It's also possible, though, to imagine paperwork as an active part of the regulatory process, inasmuch as the paper materials themselves are the point of contact where individuals experience and enact or contest the regulating authority (Caplan and Torpey 2001; Gordillo 2006; Heyman 2004). Here I want to suggest that those material qualities are constitutive of the scalar relationships that make up certification. That is, there is some property in the documentation that certification requires that creates and stabilizes particular kinds of connections between actors in such a way as to define a regional and eventually a global scale. As Tsing puts it, "Motion can be imagined as creating a global regime, whether of freedom or governmentality, only to the extent that motion is imagined as *loosening* the grip of local practice. Yet the closer we examine social and geographical mobility, the less useful it seems to think in terms of *overcoming* such friction" (Tsing 2005, 214, emphasis added). In other words, it is precisely those moments of signing forms, of holding the flow of materials accountable to local situations, that generates the structures of globalization. I refer in particular to the document known as "chain-of-custody" certification. As I noted in Chapter 2, chain-of-custody certification safeguards the purity of

Certification and the Politics of Scale   105

the commodity stream bringing sustainably produced timber to the market. In doing so, it both documents and defines a network of connection. What makes this network a particularly scale-defining moment is the particular and local quality of each connection in the chain. That is to say, the friction of specific and contingent interactions at each stage of the process, the active quality of the paperwork, is generative and productive rather than limiting and constraining. Consider my conversation with the FSC liaison at Masisa Ltd.:

> M: Well, then we'd, on the map, let's say, this is one property, this is Santa Olga, this is a property that we have in Panguipulli, that's our favorite property, because it's super well-organized, it works really well. So, as loggers we think of 'stands' as units of management. They're places that are physically together, continuous, continuous fabric, where we're going to use the same management. For example, I'm going to go there and I'm going to prune, or I'm going to harvest, or fertilize, or whatever . . . [indicates the planting regimens of various properties]. This year we're going to plant all the rest of the property. Then, for example, here we have a stand that is a management unit that's mixed raulí, and this stand has such-and-such an area, and blah blah blah.
> A: Okay, so is there a document that says "this product was made on this property," or . . .
> M: Sure, "in the property Santa Olga from such-and-such a stand." So this level of control for us is really great. That's why it's so important to have a base of really good data, supported by a good plan, because operations are like that: I say, "What are we going to cut? We'll cut this right here. Ah, Stand 22, okay, that has this area, planted in exactly this year, this species."
> A: And if you cut a tree, is it marked with the number of the stand?
> M: Just today we were thinking about doing that. Because we have on some properties were there's a lot of operations together, I don't know, they're cutting here and they're cutting there, and they're cutting here, that's the same species, that it's highly probably that the people will get confused. That's how we're thinking about marking each slab with the stand number. But in general, because our land is special, we have tiny operations that generally don't mix. Then we know pretty well that there's wood that's coming from here and other wood that's coming from there. One operation at a time, there's not really any problem.
> A: So if there's a truck, you could say, "Everything on this truck comes from . . ."
> M: Exactly, so it's not a big problem yet. In other properties, that now we're going to have strong operations in the summer, that's going to happen, so that's why today we're talking about marking each

slab, because we were seeing that the issue could get complicated for us.

A: So, a truck comes from some stand, with its number, and then where does it go, how do you document the next step?

M: For example, a track that carries native pulp [i.e. pulpwood from native species]. It comes here to Santa Olga. There's some second growth of beech, so here it's been thinned; we take some trees, leave them at the edge of the road, the truck comes and throws them and off to the plant and dumps it, that quickly. And they bring it with a legal document, guías de despacho [invoice], and in the invoice it says "such-and-such a truck, such-and-such license, such-and-such driver brought it." What's it bring? Tells you the species, if it's metro ruma [cubic meter] . . . so in this case it's Luciana, that one is metro ruma, such-and-such species, from such-and-such stand. All those details noted in our invoice, and that goes to the destination. The destination receives it.

A: Someone from the plant takes it?

M: Of course, and the guy from the plant receives it. With this invoice he sees, okay, so many meters come, such-and-such a volume of the species, okay. And since it's native it comes with its form from CONAF and there you go. They receive and give us another form for having received, and we have those two papers to track it. We control [track] what our dispatcher says went out and we also track what the plant receives. So that way we can know if, say, we've lost volume en route. In general there's no problem. The volume is almost the same, but eventually if someone were doing something weird, that's how we'd detect it. Right? These are our tracking forms. And this invoice, an associate downstairs takes it and enters it in the system, that's the same system that accounting uses. Then we have pretty good tracking of what each management unit, each stand, is producing.

Analyzing this conversation, my mind initially hung up on the linguistic accident of the term *control*. In Chilean usage *control* is a verb meaning to track or monitor; it also appears as a noun referring to the document or process of tracking. This bureaucratic usage exists in obvious tension with its everyday definition; to manage, to constrain, to operate, to discipline. At least one Mapuche activist I spoke to referred to the FSC as a *sistema de control* with the sinister double meaning very clearly implied.

That said, I think it's noteworthy that the workings of the chain of custody described in the conversation above do not appear to work with a constraining or disciplining logic. While the principle of certification might seem to indicate the expansion of a top-down optic, making local forest practices legible for global audiences, this conversation depicts a practice that makes forestry legible to foresters themselves. For the *ingeniero* with whom I spoke, participating in chain of custody certification didn't mean subjecting the company to the control (or *control*) of the FSC or of Northern consumers, but rather extending the company's locus of control throughout the network by which their products travel. If such a perspective is not a top-down model of legibility and control, it is certainly not bottom-up either; as I describe in Chapter 3, the plantation workers and neighboring indigenous communities certainly do not experience certification as an empowering tool. If we're looking to spatialize the metaphor of the chain-of-custody more than it already is, it might better be understood as taking place *through* rather than top-down or bottom-up.

I spent one cold and rainy morning talking about the chain-of-custody as I tromped along behind a Masisa crew foreman while he conducted his weekly inspection. I told him about the training I'd attended at Forestal Rio Cruces (see Chapter 3), and how the field workers seemed to have a very different set of ideas about what certification was than the FSC's board members did. He grumbled for a minute about environmentalists who never have to work and so on, then stopped me in front of a stand of *robles*.

> We'll cut these in the next round of harvests. This is good forestry—they're tall and healthy, the slope is correct, we're so many meters from the stream, there's no *huemules*[3] here. Now with FSC, I fill out the form and it goes along with the truck. So you know that yes, I'm doing good forestry. And organize it, so we always know what to cut, and when to cut, and where to cut. Other loggers, maybe they don't do the form, they just cut the trees where the trucks are, and who knows? But it's just good forestry and good business, you organize your work and you know what you're doing. Here, we follow the law, and now we follow the FSC, and so everything is very clear.

It seemed that for this foreman, at any rate, the documentation requirements for the FSC primarily represented another layer of administration.

Since the relationship between the plantation worker, the truck driver, the timber company, the national environmental organization, and the North American consumer is precisely defined by the activity of the network, the assumptions about verticality are not so clear. A significant disconnect appears between what the chain-of-custody can mean in Northern and Southern settings—or consumption and production settings, more precisely. From the consumer side, the chain-of-custody exists to constrain the "contamination" of the wood from uncertified sources, to "segregate" good wood from bad (Ozinga 2001). From the producer side, it facilitates the smooth and reliable flow of saleable product. Even as meanings fail to meet up, the commodity continues to flow, and the producer/consumer relationship is tightened. Tsing (2005) describes these dynamics as collaboration, by which she means a relationship of mutual interaction between parties across cultural or political difference that is not predicated on eliminating or bridging that difference. Without mutual understanding, without agreement on terms or goals, actors can still slip into formation with the lines of a network,

108 *Certification and the Politics of Scale*

and channel their efforts toward the production of a result that can only be seen as a network effect. Trees are grown and cut, and arrive in the Home Depot with an FSC seal of approval. Even if the green consumer and the plantation worker share no language for discussing the certification project, each play a role that is eventually written onto the landscape and the market.

The disjuncture of meaning becomes a scalar issue because of the nature of networks. Latour often points out (i.e. 1993) that one can follow the network of a scientific fact from one end to the other without ever crossing the line that separates human from animal, human from machine, or especially for our case, local from global. The meaning of those categories can only be incorporated after the fact, as a result of the associations defined in the network, rather than the cause. In the chain-of-custody example, when does a given chunk of wood leave the local sphere and enter the national, or the global? Only by imagining the plantation worker or the green consumer in relation to each other in the context of this particular network does the idea of each as 'local' or 'global' begin to make sense.

GENERIFICATION

There are other gaps that appear in the system of certification and standardization, and in some cases they do fall along lines that resemble what we might call 'top-down' or global-centric. Even then, though, the path that these (mis)understandings take is not necessarily a predictable one.

Early on in my fieldwork, I was trying to get a sense of the different versions the standards had passed through on their way towards a final format. I wondered especially, what was the starting point? I learned that in most cases, countries with new FSC initiatives and no national standards yet ratified typically use a 'generic' form of the standards, which essentially consists of FSC-International's 10 Principles and Criteria as interpreted by the auditor conducting a given certification. The purpose of the national initiative is to create a uniquely national particular version of the international standards, but certification can and did proceed in the interim. When FSC-Chile was formed, the task of writing the first draft of the Chilean standards fell to Luis Otero, who was also the first working certifier in Chile, providing interim FSC certification under the generic international version. I spoke to Otero about drafting that first version, with a particular eye to charting his individual influence on the later, finished version. What he told me about the revision process, though, sent me in a very different direction:

AH: What kind of materials did you start with? With the standards of other countries as models, or with the generic FSC standards?
LO: When I made the standards that I proposed, I read the standards of other countries, I read what had been made in England, I read the Swedish ones, there was something in Germany, I think, and there'd been standards in Brazil. I read that and I started to write.

I made the set of standards and I sent them to everyone, trying to make them the most Chilean possible. That is, attached to Chilean reality. And then they sent, they took that and started to discuss and modify. They modified, or maybe some they erased. Then, they put in others. Then they went back to the original I think that we ended up with 50%, 60% of the original. And I always alleged . . . because they always tried to take out the Chilean references that I . . . because I always put really local things, and they were taking them out. And I always alleged that, I said, "But the standard does have to be Chilean." That is, here you have to talk about araucaria forests, talk about pine plantations, you have to talk about that and that's what they took out of the final version.
AH: Who did this?
LO: No, in the process of the discussion.
AH: And how, who was participating most in this discussion?
LO: Well, it was companies, I told you about the companies, plus CODEFF, the Agrupación [de Ingenieros para el Bosque Nativo], by way of Luis Astorga, Defensores del Bosque. These are the NGOs. And then private consultants participated too.
AH: And who were the people saying to take out the local indicators?
LO: It's what they all wanted. They felt that it should be more general, I think, because, ah . . . because maybe they felt it would be very limited to make standards that referred to one reality. And furthermore, they copied, how the standards were in, how standards had appeared in the United States and other countries. And the standards were more or less general. So they were also thinking that in Chile we had to do the same.
AH: So was that controversial, then?
LO: No, because I was the only one that defended that. I and other people, but in general everyone was in agreement about doing something that would be more general.

I hope I am not revealing any lack of preparation for this field research to say that this fact took me completely by surprise. An important aspect of my project as I had proposed it was to understand the conceptual process by which FSC-Chile made international standards into local ones. And here a central figure was telling me that in a sense, precisely the opposite had happened. So I didn't do much but stew on the idea until my second conversation with Otero. Here I was inquiring about the process by which he conducted audits for certification, and yet the issue of the 'generified' standards emerged again:

AH: Can you think of an example of problems that came up during an audit?
LO: Well, there's a lot of examples. For example, people that wanted to gather mushrooms in the forests, or people that used someplace

for recreation in the forest and didn't have permission to enter, problems of local employment, that kind of thing.

AH: And how does that fit in the standards?

LO: So, well, because what I said about the criteria coincides with what I'd seen. See, I tried to write a draft [of the standards] as broadly as possible, extensive, thinking that they would cut and reduce it. And what happened was the reverse, that they broadened it and this was a . . . it was a thing I hadn't expected. And the other thing is that I made a pretty Chilean standard. For example, with things from here, I even referred to particular species. For example, I referred to the auracaria, I referred to the alerce, I referred to the species, and all that they erased after the discussion, and they generalized it.

AH: Right, right, okay. And why?

LO: Because people believed that there had to be general things and not specifics, but I think that was a mistake.

AH: Interesting.

LO: Yeah, because I had the idea of a standard much more specific to things that already exist, problems of the Monterrey pine forests, or specific problems.

AH: I remember we discussed this the first time we talked. It interests me because I thought the idea was that the standards would be made in a Chilean format, based on adapting the international standards.

LO: That was how I understood it, that it should be the most Chilean possible.

AH: But in practice?

LO: But then the people felt that it should be the most general possible. And that happened in all the countries, and that I think is a mistake. Because you, maybe . . . Took the standard from Bolivia and what there was in Peru and they were the same, took the standard in Chile and Bolivia and there isn't much difference.

AH: So why is that?

LO: Because people like to be the most international possible, they like to be . . . They don't like to write specific things, like that, I don't know. That's my interpretation. People have this idea of the international thing, they want to be international, they don't want to be . . . I never dealt with this, really. I went to some meetings, and I suggested it, "It seems to me that the standard should be more Chilean, more concrete, more attached to reality." But it wasn't very well received, I'd say, it wasn't an idea that people shared. Heinrich Burschel [German GTZ worker and FSC member] shared it a little, but I'd say the rest of the people didn't, they didn't agree with that.

AH: And what's going to happen with this issue? Is it going to have an impact in practice, on the implementation of standards?

LO: I think it will, because basically to make a standard so general . . . Now, the standard is good, I do think we did a good job . . .

AH: Well, sure.

LO: But I would have Chileanized it more, I would have put more things from here. That is, I would refer completely to the problem of collecting hazelnuts, the problem of collecting pine nuts, the problems of the Monterrey pine forest, the problems of alerce, to these questions.

AH: And without that?

LO: Without that we're left with something much too ambiguous for my taste.

AH: And there would be more room for the certifier to—

LO: More room for the certifier, yes.

AH: And as a certifier, what do you think of that?

LO: No, I think that doing it this way allows too much liberty.

By the time that I was attending FSC meetings and tracking the interactions of the players, the decisions to which Otero refers had already been made and removed from the table. Reviewing the minutes of the previous meetings actually offers surprisingly little insight; the exchanges Otero describes indeed occurred on more than one occasion, but were apparently not memorable enough for the recording secretary to document the actual language used—passages citing specific species are slated for deletion, Otero or Burschel object, others overrule and the deletion stands. Subsequent interviews with other participants, those apparently responsible for this trajectory, are equally unclear on the subject. The executive secretary of FSC was the only other besides Otero to acknowledge the issue; he told me that the language had to be streamlined to be applicable to all possible situations. "*Si digamos algo acerca de, no sé, alerce o radiata, ¿que pasará si tendramos problemas con otro especies?*" [If we say something about, I don't know, *alerce* or Monterrey pine, what will happen if we have problems with other species?]. For him, it was a simple mechanical matter, a legalistic problem of avoiding unforeseen complications with the language.

But from Otero's perspective, "people have this idea of the international thing, they want to be more international." The strength of the FSC as a regulatory mechanism, after all, derives from its internationalism, its conjuring of the international market to reshape the national landscape. Even Mapuche activists, skeptical about the FSC itself, invoke the role of the international timber market as a determining factor in local landscapes and conflicts: "In the environmental aspect, I believe that conscience has come about thanks to the current international market demanding certified products. The conflict isn't only a rural situation, rather it's significant at the level of the international market."

The appeal to the international market is a discursive one, an attempt to apply scale frames (Kurtz 2003) to a political economy with particular ends.

What makes this episode particularly interesting is the way those frames, once articulated, are inscribed in standards and re-broadcast. Chile's FSC standards are now more generic, less particular than they might have been. Perhaps this provides them with the regulatory flexibility the executive secretary mentions; at the same time, it calls into question the very purpose of a national FSC board and national standards. Moreover, as Otero points out, the lack of species-level specificity leaves more discretion in the hands of the auditor. Redistributing the authority of certification in this fashion could have potent effects. Even before the national standards had been defined in Chile, one auditor (Institut für Marktöcologie) had been suspended for issuing 'fraudulent' certifications. More generally, the elision of particularity suggests that the association of given actors or projects to particular scales is subject to change without notice, according to reasons that may not be clear to anyone. It is in this space of uncertainty that the politics of scale plays out its unintended consequences.

THE STATE

The state is and is not present in Forest Stewardship Council certification. By definition, a voluntary certification program is categorically distinct from state-based enforceable legal mechanisms; some analysts use the term "non-state market-driven" regulation (Cashore 2002; Cashore et al. 2004). The FSC makes a point of policing that boundary and defending their independent status. Connections made between local communities or organizations and transnational activist bodies are sometimes described as "scale-jumping" practices (Herod and Wright 2002; McCarthy 2005), in that they attempt an end-run around the influence of the state. In practice, though, the relationship between certification programs, the state, and individual state officials appears to be quite a bit more complex. I close this chapter with a discussion of the ways in which the Chilean FSC project complicates the "non-state" claim, and consider what this tension means for the project of person-making in a global network of ethical consumerism.

The locus of authority in a non-state market driven policy system (NSMD) is presumed to be categorically distinct from the government. Cashore defines a typology that ranges from government in the classic sense; through "shared public/private governance" in which the state is still (implicitly or explicitly) the ultimate source of authority; to NSMD governance, in which the ultimate source of authority resides with market transactions (Cashore 2002). This is the framework that FSC-Chile members implicitly invoked when they described their participation in certification projects as a response to Chile's continuing failure to develop a native forest law.[4] In nearly every interview I conducted, someone would refer to the Ley de Bosque Nativo with a sort of pointedly rueful chuckle that became very familiar. The implication seemed to be that if the state was so clearly incapable of regulating the harvest of native forests, despite everyone's best efforts, then that goal would best be pursued outside the state. The argument for forest certification as a sort of "regulation in the absence of regulation" functions at the global scale as well—in the absence of a global forest convention signed by states (Lipschutz 2001), global forest activists have at times cast about for alternatives (Vanclay and Nichols 2005; Bernstein and Cashore 2004).

Cashore cautions against treating NSMDs as examples of "negative state participation," that is, of states deliberately devolving authority and responsibility for regulatory tasks to non-state actors. I would argue, however, that in some cases that is exactly what is taking place. Certain examples close to home seem to make the case, as in Clinton-era welfare reform and 'volunteerism' (Hayward 1998), or the Bush administration's Clear Skies Initiative (Lutter and Shogren 2004). The Chilean scenario makes a strong case for the persistent influence of the state in NSMDs at an important structural level.

One way in which this becomes clear is the part played by CONAF *[Corporación Nacional Forestal]* in giving shape to FSC standards in Chile. Essentially, CONAF regulations provide the baseline for certification. In large part, this is the logical outcome of FSC-International's first Principal: "Forest management shall respect all applicable laws of the country in which they occur, and international treaties and agreements to which the country is a signatory." CONAF is the institutional embodiment of all forest-related laws in Chile, and the body charged with their enforcement. It's natural, then, that the unavoidable first step in certification according to the standards of FSC-Chile be that all relevant CONAF regulations be observed, and that properties be able to present a *plan de ordenación*[5] approved by CONAF. On the one hand, this seems to reinforce the point above, that certification is taking on the mantle of enforcing laws that CONAF should but, underfunded and understaffed, cannot. The *plan de ordenación* is more than a first step or a minimum threshold in certification, though, and in fact appears throughout the final version of the standards. Each of the 10 Principles has a list of associated criteria, each with a range of indicators; in most cases, the indicators are accompanied by *verificadores*, or verifiers—the form of evidence that will document the compliance with each indicator. Examples include documents (titles, contracts) and processes (consultations with interest groups, training events); of more than 200 indicators included in the standards, nearly half were to be verified by the existence of a *plan de ordenación* approved by CONAF. About three pages into the standards, *plan de ordenación* is replaced by the abbreviation *PO*. The ubiquity of the CONAF document suggests that the state, in the form of CONAF, is present as more than a baseline for certification.[6] In a certain sense this state proxy is also the authoritative voice in documenting compliance, indeed in defining what it means to comply with standards at all.

FSC-Chile has been fairly explicit at times about its perspective on the active role of the Chilean state in certification, as well. Particularly in regards to what many in the forestry sector call "the social question," FSC-Chile has cast the state as the broker for all forms of development initiative:

> Timber companies are the most important economic actors in the timber regions because they possess important natural resources and generate production, work, income, taxes and other actions of local impact. For this reason, local development will depend to a large extent on them. Nonetheless, to make local development happen, it is necessary to work together with all of the actors including rural communities, municipalities, organizations, institutions and others in which each can make their own contribution and participate in decisions. *The municipality, as the democratically elected local authority, must realize the coordination of all local development.* (Astorga 2004, 49, emphasis added)

It is not clear that timber companies are willing to authorize local municipalities to realize the coordination of all local development. Perhaps this is why Astorga felt it necessary to emphasize this point; in practice, timber companies have tended to internalize the developmental demands made on them by FSC certification. That is, several companies have erected school buildings for communities neighboring their properties, or provided their own series of agricultural technology outreach and *capacitación* events (Catalán 1999). Taking these obligations on as projects allows the companies to capitalize on the goodwill generated with little risk; letting even municipal authorities dictate the development measures to be taken opens companies to criticism and accountability they might prefer to avoid. Even less likely to accept such an allocation of authority are the Mapuche communities, already feeling marginalized by state-based development planning (Durston and Duhart 2003). One of the few themes widely shared by Mapuche organizations and community leaders is the principle of *autodesarrollo* [self-development] or *"nuestro propio desarrollo"* [our own development].

That Mapuche communities and activists should hesitate to accept the state as the broker of local development should not be surprising. The most visible form of state-sponsored development in Mapuche territories is the plantation sector, the beneficiary of decades of government subsidies. Plantations have become a dominant force on the Chilean landscape as the result of a sort of perfect storm of political and ecological factors in which the state is unavoidably implicated. The socialist administrations of the 1960s legislated hefty government subsidies for farmers growing trees (Camus 2005). During the military government of 1973–1989, the Pinochet dictatorship privatized state resources and eliminated most forms of government social spending. But the tree-planting subsidies continued, so that Chile's largest companies were being paid vast sums to cut down native forest and plant profitable pine plantations. Chile's reputation as an economic miracle, a success story of privatization and neoliberal adjustment, is due in large part to the state-subsidized timber sector, which remained profitable during a period of transition and economic instability (Camus 2005; Clapp 1995b, 1998b). Despite the free-market ideology of the dictatorship and its US-based supporters, those timber profits were massively underwritten by state subsidies. Even to the extent that the FSC is successful in isolating itself from direct state influence, all forestry in Chile has to reckon with the legacy of subsidies and the plantation economy.

Of course, timber subsidies were hardly the only thing the interventionist state has contributed to the plantation economy. More insidious, and less quantifiable, is the role of the Chilean state in managing labor and indigenous conflicts. In fact, the Chilean state has a 100-year history of managing labor markets for the timber industry by breaking up *campesino* cooperatives, criminalizing organized labor, and even building prison labor camps for indentured forest workers (Klubock 2004, 2006). The state has been displacing the Mapuche since long before the emergence of the modern timber industry, but the relationship between the two has been well documented in the late 19th and 20th centuries (Bengoa 2000a; Camus 2005; Pinto 2003). The process reached a high point during the military dictatorship, when the Pinochet government single-handedly rolled back agrarian reform, and took literally millions of hectares of land held by *campesino* cooperatives and Mapuche communities—and handed them over to large corporate landowners (Bengoa 1999; Frias 2003).

To a certain extent, the FSC manages to isolate its activities from the colonial process with a sort of historical firewall. FSC-International came into being in 1994; as a result, all of the Principles and Criteria worldwide are essentially 'backdated' to that point.[7] For example, the FSC will not certify timber from a property converted from native forest to plantation since 1994. Similarly, territorial conflicts arising since 1994 are theoretical barriers to certification in Chile; since that date happens to fall after the end of the military dictatorship, many if not most potential conflicts are essentially grandfathered out of FSC-Chile's jurisdiction.

However, the state continues to intervene actively in territorial conflicts between Mapuche communities and timber companies, and discursively and legally defines Mapuche political activists as 'terrorists' (Araya 2003; Frias 2003; Richards 2007; Seguel 2001). By casting activism as terrorism (rather than political activity or petty crime), the state makes its role in timber conflicts explicit. This is the argument that most compromises the "non-state" model—even when 'hollowed-out,' the state is never really absent (see Friedman 2004; Robinson 2004). The persistent presence of the Chilean state in an ostensibly non-state context speaks directly to the heart of Swyngedouw's (2000) argument about authoritarian state violence; rather than disappear in the new environment of transnational activism and global connections, state violence is redeployed, reimagined, and obscured. But the state remains an overwhelming presence in its original role as territorial power,[8] the builder of borders and defender of private property. Programs for genuinely non-state governance are incoherent in the context of this enormous monopoly on territory.

CONCLUSIONS: ACHIEVING UNIVERSALS

Perhaps the most successful form of non-state market driven regulation—depending how we define success—is organic agriculture. I have used organic certification as a comparative case throughout this project and in casual conversation, in no small part because it is so familiar to so many people. The familiarity of organic is a relatively new phenomenon, though, and it comes with some serious implications for the practice as a whole. Geographer Julie Guthman has taken on the relationship between the political economy of organic and the meanings that it circulates (its value and its values, if you will) in a way that illustrates some very useful points on forest certification as well.

To begin with, US federal organic standards (like FSC standards) are scale-free in the explicit sense; there is no defined limit to the size of the operation to be certified. Other factors being equal, a 1,000-acre monocrop farm is as certifiable as the smallest and most self-consciously 'green' operation. In the absence of scalar constraints, economies of scale inevitably introduce a cost pressure that works at cross-purposes with the values commonly associated with organic agriculture. In Guthman's view (1998, 2002), certified organic agricultural responding to market pressure eventually becomes distinguishable from conventional agriculture only by the inputs used—liquefied kelp instead of chemical nitrogen fertilizer, for example (see also Allen and Kovach 2000). The California organic production systems that make up Guthman's case study are shaped by a particular set of political economic drivers; in this case, the vertical integration of growers and the staggering cost of real estate in California. The case of FSC-Chile, as I've argued here, has its own set of scalar problems. On the one hand is the land problem, understood as the legacy of colonialism and dictatorship in relation to indigenous territory; on the other is Chile's place in the world economy relative to the buyers of its certified product.

For a Marxist like Guthman, these political economic problems of scale are inevitabilities, qualities and effects inherent in operations that channel power and finance in certain patterns. Without downplaying the pressure of markets on a land-constrained production system, I want to reiterate that discursive aspects of scale are equally at play in the case of FSC-Chile. In particular, the international market is ostensibly the driver of social change in a system like forest certification—but as I note in Chapter 2, Chile's timber companies are neither receiving nor expecting a price premium for certification. For timber companies, environmentalists, and social activists alike, however, the international market is indubitably present as a discursive force; the market provides the "significant elsewhere" (Lowe 2005) of forest certification in Chile.

"The Market," then, in and of itself (as they say in Chile, *'en mayasculo,'* [in capital letters] to refer to its singularity) is a universal, an ideal type, an invisible hand. It is a set of forces imagined to obtain equally in all places and times, like the pull of gravity. Anthropologists as scholars of difference tend to pick at constructions alleging universal value in this fashion, and offer up ethnographies and case studies of economic activity taking place outside of, beneath, or in opposition to the forms and ideals economists know as 'market forces' (Bohannon and Bohannon 1968; Carrier 1995; Godelier 1977; Graeber 2001; Taussig 1980; Wolf 1982). We are not inclined to accept economic forms, as economists know them, as any kind of 'human nature.' As Carrier describes (1997), neoclassical economists make market forces *appear* universal through a two-stage cleansing process. First their theorizing removes traces of non-rational or excessively 'cultural' behavior from recognizably Western exchange proceedings. Then the economic practices of cultural 'others' are re-read according to the principles of rational economic behavior, cost-benefit analysis, and similar concepts. Once the invisible hand is observable everywhere, it is easier to craft policies that empower its movements.

Forest certification, as an NSMD form of environmental regulation, was designed to link the universal forces of the international market with another globe-traveling universal, nature. The laws of nature, by their very definition, are everywhere the same. Scientists, environmentalists, and politicians all benefit in their own fashion from imagining nature as a global equalizer (Williams 1980; Cronon 1996; Ingold 1993). But like the Market as Carrier describes it, universal nature is not born, but made. To complicate the picture further, universal nature is made by actors self-consciously manipulating discourses of universality and locality to achieve political ends. In Choy's terms:

> Our concepts of 'universality' and 'particularity' . . . reach a point of analytic failure when confronted with transnational environmental politics. In such environmental arenas, the very concepts of 'universality' and 'particularity' are in a constant process of self-conscious deployment, production, and articulation. This is nowhere more plain than in environmental controversies in postcolonial states, where arguments for and against the universalism of both political commitments—such as 'environmentalism'—and knowledge forms—such as 'science' and 'local knowledge'—are always already in play. (Choy 2005, 6)

Tsing (2005) describes four vectors by which scientific and political action has cast global Nature in a universal frame. In the first vector, the foundation of botanical classification played a crucial part in the birth of modern science; it defined a universal nature in the tension between a systematic logic for rationally ordering plant diversity, and the empirical collection of diverse individual organisms. In the second, John Muir provided the religious rhetoric that shaped the emergence of North American nature protection; Muir's protestant theology cast Nature as an omnipotent God, to be approached through individual encounter and meditation. In the third vector, global climate modelers devise systems for integrating data at all scales into nesting hierarchies that "simplify and reduce the social and natural world to geophysical laws. In the process, they develop a globe that is unified, neutral, and understandable through the collection and manipulation of information" (2005, 102). In the fourth, the International Tropical Timber Organization attempted to promote global standards for the "sustainable management" of tropical forests that would satisfy the agendas of timber companies and environmental advocates alike. With representation divided evenly between "producing countries" and "consuming countries," and votes within each block assigned according to market share, the resulting International Tropical Timber Agreements of 1994 and 2006 largely consisted of empty definitions and ineffective policies (see also Poore 2003). Regardless of their success in their own terms, these globe-making projects achieved some coherence by coupling a universal notion of nature with reason, with God, with science, and with commerce.

Forest certification joins this lineage inasmuch as it intends to hitch the protection of nature to the functioning of the free market. But the coupling is not a smooth or automatic process—as in Tsing's examples, the framework for imagining the universal must be assembled out of existing pieces. As Jameson says, "the universal is not something under which you range the particular as a mere type" (2002, 182, cited in Tsing 2005). It is perhaps better described as the process of translation, of reaching for a language to enable connection across difference. When environmentalists appeal to universal values to support their attempts to protect a species or a landscape, they seem to write out alternative perspectives that may not share those values after all (Redford and Stearman 1993; Chapin 2004). But sometimes those stories have legs—sometimes they circulate in ways not anticipated in the moment of telling. In a particularly telling example, Tsing recounts a unique re-purposing of the Chico Mendes[9] story in the hands of an Indonesian environmental activist. Attempting to encourage rural Meratus villagers to organize themselves against logging interests, the activist recounted Mendes's struggle and the successful creation of Amazonian extractive reserves, but ended the story with an odd hybridization:

> The woman telling the story drives the message home. 'When they came to cut down the forest,' she says, 'the women came out and hugged the trees.' She reaches her arms out, miming the hugging of trees, and the other woman activist joins her, showing people how the women hugged the trees. I believe we have moved from Chico to Chipko, from the rubber tappers of the Amazon to the foothills of the Himalayas where women came out to protect the forest by hugging the trees, thus the name Chipko, hugging. (Tsing 2005, 230)

Lifted out of their contexts, these narratives provide new motivating values to activists seeking grounds for collaboration.

In a more concrete example, medical sociologists Timmermans and Berg document the particular strategies required to create a universally accepted cardiopulmonary resuscitation protocol (1997, 2003). They propose the term *local universals* to incorporate the ambiguities of actor and mechanism as found in Tsing's vectors to the universal that I describe above: For Timmermans and Berg, "local universality emphasizes that universality always rests on real-time work, and emerges from localized processes of negotiations and pre-existing institutional, infrastructural and material relations. 'Universality,' here, has become a non-transcendental term—no longer implying a rupture with the 'local,' but transforming and emerging in and through it" (1997, 275). The standardized protocols in question are fascinating in their power to obtain effective action across a range of actors—CPR must be equally usable by physicians and laypeople, in any conceivable location or situation. The power and universality of the CPR protocol, however, is a very recent innovation—the combination of closed-chest cardiac massage and mouth-to-mouth ventilation was not tested or promoted widely until 1977. Prior to that time, a wide variety of ventilation and cardiac stimulation techniques were used in different countries and different settings (military, emergency room, clinic). The American Heart Association in coalition with the International Red Cross and a number of other institutional actors collected and analyzed thousands of records based on existing methods for treating near-dead patients. The results were laboriously synthesized in professional meetings and international summits, eventually yielding CPR as we know it. But even in their current, stabilized, and well-known form, CPR protocols are subject to contestation by individual practitioners—emergency personnel in particular alter protocol 'on the fly' according to their experience and subtleties of the crisis situation. CPR protocols, in other words, are the result of contingent lashups of various methods, assessed and standardized by institutions, and employed or modified by individuals with a multitude of influences. Timmermans and Berg describe the new, universally adopted life-saving protocols as "technoscientific scripts which crystallize multiple trajectories. In the process of obtaining local universality . . . protocols feed off previous standards and practices . . . protocols function through the distributed work of a multitude of actors . . . in this process, protocols themselves are necessarily changed and partially reappropriated" (Timmermans and Berg 1997, 273). This 'local universality' is the middle-range outcome of the project that attempts to transform a messy set of practices into an ideal form; the result is patchy rather than uniform, and shifting rather than perfectly fixed. It is precisely these qualities that make protocols flexible enough to remain intact in a diversity of settings; not Latour's 'immutable mobile' (1987), but rather just-mutable-enough.

The point that I'd like to make in closing this chapter is that the ontological and discursive aspects of scalar politics can't easily be disentangled in the

Chilean case. Not only do discursive scalar claims have material outcomes (Kurtz 2003), the two are mutually reinforcing in the political sphere. In Chilean certification, parties to a negotiating process define themselves and their interests according to scaled models that are in turn part and parcel of political discourse. Thus, as Dunn notes below, the micropolitics are constitutive of the macro:

> Seeing standards not just as a bounded set of global rules but as regulatory mechanisms which enter into specific contexts opens up the possibility of unexpected and unforeseeable results of economic transformation. Peasant resistance to standards in Poland, for example, suggests that standardization, as a governance technique of neoliberal capitalism, elicits a conflicting complementarity such as that engendered by state socialism. The ways in which Polish producers accept standards, resist them, or circumvent regulations will help to shape the way Poland is integrated into regional and global economies . . . the rather small-scale actions of rural people have potentially large-scale consequences. (Dunn 2003, 1508)

As I've attempted to demonstrate in this chapter, it is not only the small and rural actors who are beholden to scalar models of political action; the 'generification' issue and the problematic role of the state illustrate higher-order scalar complications that accompany the appearance of standardization as a regulatory form. If Dunn is right about the potentially enormous impact of small-scale actors, though, then the stakes in imaging Chilean forest certification as a global project may be larger than at first they appear. This is the trajectory the FSC aims for when it brings parties together to negotiate standards, and it requires a reimagination of what the universal means. What happens if we array this distributed, achieved universal against the other set of scales described in this chapter? The politics of scale become, perhaps messier, but also more open to possibility. Things that don't make sense, or seem just inappropriate (like the generification of standards, or the reification of the small proprietor) take on a shape and consistency when considered as small translations. Perhaps we should see these incongruencies as preliminary steps in the crafting of a local universal, a boundary object that might hold the diverse actors of forest certification in a rough, temporary coherence.

NOTES

1. Chile has a higher literacy rate than much of Latin America, but still shows a sharp urban–rural divide on the issue, and low-income small forest proprietors, even where literate, are unlikely to purchase an illustrated manual at 35 US dollars.
2. Apologies to Latour (1987).

3. The Hippocamelus bisulcus or Andean deer is one of Chile's national animals, endangered, and very rare. The foreman's use of the *huemul* is ironic: Of course there weren't any in this patch of managed forest.
4. The *Ley de Bosque Nativo* finally passed in December 2007, after 15 years of debate and compromise—the longest legislative struggle in Chile's history. For more, see Lara et al. (2003) and RENACE (2001).
5. Even the term itself is evidence of CONAF's pervasiveness: The generic term *plan de manejo* [management plan] appears in other Spanish-language standards and in early drafts of FSC-Chile standards. *Plan de ordenación* is the term used for CONAF's official documentation, and was adopted into FSC standards before their finalization.
6. See Meidinger (2001) for a similar analysis of forestry in the US; also see Newsome et al. 2005.
7. The use of legislative dates as benchmarks in environmental regulation is a fascinating topic beyond the scope of the present discussion. A not-entirely tangential example is Chile's 1970 Alerce Law, which bans the cutting of living *alerce* trees. Standing *alerces* deemed to have died before the bill was signed into law are still legal to cut and trade. As a result, a complex cycle of trade and enforcement has emerged in which forest owners mysteriously discover 'long dead' *alerce* on their property, while conservation scientists devise forensic methods to date the deaths of *alerce* found in the market, including taking core-samples from furniture (Wolodarsky and Lara 2005).
8. For an excellent overview of the Chilean state as a territorial power vis-à-vis the Mapuche, see Bryan (2001).
9. Leader of the Brazilian rubber-tappers union, famous for mobilizing a grassroots defense of Amazonian rainforest that framed the forest as a place of livelihood rather than pristine wilderness (Mendes and Gross 1989).

6   Conclusion

Quien no conoce el bosque chileno no conoce este planeta.
—Pablo Neruda, "*Confieso que he vivido,*" 1974

QUIEN NO CONOCE EL BOSQUE

I open these closing comments with this quote from Chile's canonized poet: "Whoever doesn't know the Chilean forest doesn't know this planet," and I should say that I have had what could be called a love–hate relationship with this piece of language for quite some time. When I first read Neruda's *Memorias* I was thrilled to find such a pithy piece of language that touched on my intellectual concerns so well, and I put it in all of my grant applications and conference presentations. Once I spent some more time in Chile, I began to feel a bit silly about my attachment to Neruda's poem, as I realized that nearly every environmentalist organization, tourist operation, and chamber of commerce in the country was citing it too. On the other hand, if the poem has been commodified to the point of meaninglessness, that makes it all the *more* pertinent for a study of an international green seal of approval for sustainable forestry. In any case I find myself unable to resist the multiplicitous quality of knowledges that Neruda implies—or that we can infer, if the reader will allow me a free hand with interpretation. In the first instance, we find the multiple ways of 'knowing' the forest: As a hotspot of biodiversity, as a landscape of indigenous cultural heritage, as a productive economic unit. In the second, Neruda cites a knowledge of the Chilean forest as a prerequisite to 'knowing' the planet. What can this possibly mean? With the right attitude, we can read this claim under the sign of friction, of emergent and contingent concepts of the global and the universal.

"Friction," again, is Anna Tsing's metaphor for the material particularities of global connection: "emergent cultural forms—including forest destruction and environmental advocacy—are persistent but unpredictable effects of global encounters across difference" (2005, 3). In the preceding chapters I have explored one such space of encounter—the virtual negotiating table where loggers, environmentalists, and other affiliates of the Forest Stewardship Council (including one lone indigenous Mapuche representative) came together to hammer out the criteria and indicators that would define sustainable forestry in Chile. Labeling schemes like FSC are generally described as "values-based" and "stakeholder-driven," but the space of encounter at the negotiating table, and the values represented there, are neither transparent nor pre-determined. Rather, they are the dependent outcome, shaped by power relations and institutional constraints. In other words, the 'goodness' of 'good wood' is made, not grown. As a result, we can speak of certified wood and its qualities as cultural artifacts, even network effects. I will have to leave for other researchers any assessment of FSC-Chile in terms of the biodiversity it conserves or the sustainable income for local business it generates. But the conversations and interactions documented herein do offer suggestive materials for evaluating the meanings associated with certification as it defines itself into being in Chile, and for speculating about its ethical implications and possibilities.

GETTING WITH THE FETISH

In a previous chapter I referred to the widely cited comment by geographer David Harvey that "The grapes that sit upon supermarket shelves are mute; we cannot see the fingerprints of exploitation upon them or tell immediately what part of the world they are from. We can, by further inquiry, lift the veil on this geographical and social ignorance and make ourselves aware of these issues" (1990, 423). Harvey is rightly regarded as the founding figure in a new wave of Marxist social science; indeed, he has described his own project as the rigorous application of the original texts of Marx and Engels and "very little else" (see Castree and Gregory 2006). Here, Harvey is bringing the issue of commodity fetishism directly from Marx into the enormously complex realm of modern international circuits of provision. Even the most orthodox Marxism acknowledges the liminal qualities of the commodity form; ascribing powers of social origin to the imaginary agency of the inanimate. Significantly, the commodity appears as the vehicle for the accumulation of capital and economic power in anthropological work of Mintz, Wolf, and so on. But at about the same time, historically speaking, anthropologists began rethinking the place of the commodity form in culture.

Taussig's *The Devil and Commodity Fetishism in South America* (1980) drew harsh criticism for, among other things, its highly idiosyncratic re-reading of other anthropologists' field data. His central principle, though, the idea that commodities and the commodity form could be imagined radically differently in a different cultural context, became enormously influential. The Colombian *campesinos* and Bolivian miners in Taussig's account invested money and the commodities it could buy with demonic meanings completely at odds with elite practice or neoclassical economics. At the same time, Taussig reflects the money ideologies of his subaltern subjects against the taken-for-granted fetishizations of capital in contemporary Western society, such as money that grows, or climbing interest rates. Both examples demonstrate the loaded cultural content of money and commodity forms imagined in economic theory to be smooth, frictionless, universal solvents. Importantly, Taussig demonstrates that fetishized commodities can represent more than false consciousness on the part of oppressed classes; fetishism can also be a potent act of political resistance.

If Taussig's book was an idiosyncratic rethinking of Marx, Appadurai (1986) turned the insight into a blueprint for a new way of thinking about commodities in anthropology. In calling for a sort of "methodological fetishism," Appadurai argued that "we have to follow the things themselves, for their meanings are inscribed in their forms, their uses, their trajectories ... from a theoretical point of view human actors encode things with significance, from a methodological point of view it is the things-in-motion that illuminate their human and social context" (1986, 5). A vertical rather than a horizontal slice, so to speak, allows the analyst to define biographies of commodities. A biography is open to meanings and possibilities that may not look like the ideal forms in the models of neoclassical or Marxist political economies.

That book, *The Social Life of Things* (Appadurai, ed. 1986), could be considered the founding document of contemporary material culture studies. With these interventions, along with the renaissance of critical geography in the 1990s, the 'cultures of consumption' thesis came into its own (i.e. Miller 1995a, 2003; Jackson 1999; Cook 1994; Cook and Crang 1996; Sayer 2003, and etc.; see also Castree 2001). The notion that consumers of commodities were not passive receptacles for the detritus of capital, but rather active agents generating identity, clearly resonated with a number of political and academic imperatives. Taken as a whole, the ethnographic literature on commodities and their cultural signification makes Harvey's mute grapes appear somewhat problematic. In particular, studies of the re-signification of commodities by ethnic minorities in opposition to dominant cultural values suggest that casting consumers of fetishized commodities as victims of false consciousness not only lacks nuance, but also imposes an ethnocentric model of value and political agency.

That said, the sort of scholarship that emerged from this synthesis has been characterized at times as no more than a celebration of shopping (Maddox 1997). Wilk suggests that even if we can no longer advocate for uncovering the authentic nature of produced commodities behind the veil of fetishism, social science still has a normative duty to fulfill. It is a fine line that Wilk is walking here, and he is at pains to avoid the kind of rhetoric (both conservative and progressive or Marxist) that condemns consumption as an inauthentic practice of the bourgeoisie or a self-destructive habit of the marginal. Grasping the horns of that dilemma, Wilk argues that the specter of global environmental crisis makes it impossible (or at least unwise) to consider consumer habits as cultural practice in isolation from their material impacts on a global scale (Wilk 2001). In this sense we can read him as echoing the concerns expressed by Roy Rappaport (1993) about contemporary consumer economies, and the social sciences that analyze them, as complicit in the "subordination of the fundamental to the contingent and instrumental" (1993, 299). Without surrendering our right to question the construction and ascription of what's fundamental and what's contingent here, we need to respond to the call to imagine social analysis of economic and environmental practice in an ethical, or indeed a moral, as well as ecological, context.

This is not a simple question. I have argued throughout this work that the ethical content of FSC certification, its goodness and greenness as it were, are network effects.[1] If we understand standards and certification as a project of distributed environmentality (*pace* Agrawal 2005), we can say that the network simultaneously produces the good wood and the subjectivities of those actors who are party to it. So how to pin down the locus of ethical responsibility and political agency?

To begin with, "the market" for FSC-certified wood is not to be taken at face value. The efforts of economists and certification's promoters to identify a price premium for certified wood have yet to turn up a consistent value (Cashore et al. 2007; Haener and Luckert 1998; Rametsteiner and Simula 2003). This economic reality is subject to change, of course, especially given the growing global concern about climate change. The emergence of carbon markets and related market-based policy responses like REDD (Reducing Emissions from Deforestation and Degradation) could completely transform certified forestry's role in global markets. As I document in Chapter 2, parties to FSC certification in Chile approach this uncertain environment with a set of motivations that cannot be reduced to simple rent-seeking. Second, FSC standards in Chile have been defined by a negotiation process that built in a particular hierarchy of knowledge. Chapter 3 attests to the implicit and explicit inscription of the cultural politics of knowledge in standard-making. A conventional set of authoritative knowledges are normalized and privileged, while indigenous knowledge is discounted as 'politics,' and the embodied practical knowledge of forest workers simply does not register. And third, the theory and practice of setting standards for a global, ethical, or 'green' NSMD imposes a series of scalar distortions. As Goodman notes, ethical trade initiatives imply a political ecological imaginary (Goodman 2004). In Chapter 5 I argue that FSC-Chile has articulated itself into that imaginary with a collection of unexamined assumptions about scale, which place its efforts into a tenuous relationship with the ostensibly universal values of justice and sustainability that inspired certification as a project in the first place.

To make any final theoretical or normative arguments about this morass, I have adopted as a sort of slogan Taussig's incitement to "get with the fetish" (Taussig 1992). Like most of what this iconoclastic figure has contributed to anthropology, this epigram is pretty open to interpretation. I read it as an injunction to a pragmatic[2] style of ethics, one that takes on board all of the issues of shifting meaning and global circulation in imagining the ethical.

TECHNOLOGIES OF VALUE AND 'MORE WORK FOR MOTHER'

My name for this messy catch-all of knowledge and ethics in political economy is *technologies of value*. Certification is a set of abstractions, constructed by particular actors to manipulate materials and information towards certain ends. In the process, existing structures of power and knowledge are simultaneously reshaped and reinscribed. My use of technology here, which draws on existing literature on technologies of government (Agrawal 2005; Escobar 1995; Ferguson 1994; Foucault 1978, 1980, 1991; Rose 1999), is also an attempt to recruit meaningful metaphors from social or cultural studies of technoscience. In particular, I think that understanding technologies as historical artifacts can shed new light on problems of environmental regulation as an institutional project.

Brosius (1999a, 1999b, 2003) charts the progressive institutionalization of environmentalism; echoing Rappaport (1993), he expresses concern that although expanding environmental institutions may "create certain possibilities for ameliorating environmental degradation, they simultaneously preclude others" (1999a, 38). In the social history of technology, this problem is referred to as 'path dependence':

> Technologies often manifest increasing returns to adoption. The processes of learning by doing and by using . . . and the frequent focus of inventive effort on removing weak points ('reverse salients') from existing technologies, mean that the very process of adoption tends to improve the performance of the technologies that are adopted. This gives the history, especially the early history, of a technology considerable significance . . . Path-dependence means that local, short-term contingencies can exercise lasting effects. (MacKenzie and Wajcman 1999, 19)

As a technology of government like forest certification becomes a reality, it also hardens the channels by which it was constructed and delimits the possibilities for future governance projects. It is not for nothing that more than one of my informant used the exact phrase, "*desde ahora, es puro FSC acá en Chile,*" [from now on, it's pure FSC here in Chile].

Everyone understands that environmental regulation is political; understanding regulation as a *technology* allows me to highlight the ways in which that politics is built into the infrastructure of regulation itself. When Winner asks, "do artifacts have politics?" he is pointing out that agency of infrastructure: "the theory of technological politic draws attention to the momentum of large-scale sociotechnical systems, to the response of modern societies to certain technological imperatives, and to the all too common signs of the adaptation of human ends to technical means" (Winner 1999, 29). Winner defines two ways in which artifacts can have politics: (1) A specific technical procedure becomes a means for a particular group to settle an issue in their interest, and (2) certain technical systems may require or be strongly compatible with certain sociopolitical relationships. In the first case, Winner refers to urban planner Robert Moses, who reportedly designed the low overpasses on the Long Island Expressway to deliberately exclude public transit (and with it certain "undesirable" elements of the population) from the beaches and bedroom communities of Long Island. A proactive framing of the same issue is embodied in the Americans with Disabilities Act and the movement to promote technologies of accessibility. The second case, that of inherently political technologies, is typically made by comparing nuclear power (*requiring* a semi-police state) and solar power (*compatible* with decentralized democracy): "If we examine social patterns that comprise the environments of technical systems, we find certain devices and systems almost invariably linked to specific ways of organizing power and authority" (1999, 34).

Significantly, though, Winner's theory of artifactual politics emphasizes the role of deliberate and active political agency on the part of those planning or engaging with infrastructure. But a significant trend in science studies has taken issue with such a human-centric model of technopolitics (Latour 1987, 1993; Law and Mol 2001; Callon 2001); technologies do not mean or act only in the ways we intend them to. Technologies can only constrain or impose power in the way Winner describes by virtue of enrolling or interpellating the right human and non-human actors. This is an ontological issue for Latour and the others implicated in actor-network theory (Latour 2005), but it finds expression in a million quotidian ways—the kind of thing we generally understand as "Murphy's Law." The more technologies enroll diverse actors, and the more they extend to new scales of influence, the greater the unintended consequences that may emerge. In other words, the more parts there are, the more things can go wrong. Historian Edward Tenner's *Why Things Bite Back* (1996) is an entertaining (or horrifying) catalog of such possibilities, familiar and otherwise; the increased use of antibiotics has reduced infections but facilitated the evolution of 'superbugs,' air conditioning cools buildings but contributes to the heat effect of cities, computers designed to improve the performance of office work can cause carpal tunnel syndrome.

When I think of certification as a technology of value, I keep coming back to the scenario I mentioned in Chapter 2 when I introduce my project to friends or family. One conversation in particular stands out, in which a normally loquacious colleague was rendered genuinely speechless by the challenge of knowing what kind of paper to buy. We have to print things out, no matter how hard we try to use the various electronic features that are proliferating across our desktops. But is it better to buy paper made of wood from certified forests, or recycled paper made from wood of unknown provenance, or some combination, and if so how can we know the relative content, not to mention the problem of defining 'sustainable forestry' that's taken up so many pages here already? "It's so hard," he kept saying, "I wish there was a way to make it easier." It sounds like a commercial for forest certification, doesn't it? It's the kind of consumer lament that makes marketers or inventors see a niche to be filled, a service in need of provision. This is how I imagine certification as a labor-saving device, a machine to make consumer decisions about forest products easier.

But as a labor-saving device, certification is heir to a complicated history. Suchman (2007) points out the social aspects of unintended consequences that attend the development of labor-saving technologies, and the invisibilities and silences they promote: The "machine's presentation of itself as the always obliging, labor-saving device erases any evidence of the labor involved in its operation. . . . Yet as Ruth Schwartz Cowan (1983) and others since have demonstrated with respect to domestic appliances, the effectiveness of any labor-saving device both presupposes and generates new forms of human labor" (2007, 221). Cowan's *More Work for Mother* (1983) is a classic in the history of technology, and highlights all of the ironies wrapped up in the idea of a labor-saving device. On the one hand, labor is less "saved" than it is deferred to other places and actors. An electric mixer, for example, can mix dough with less muscle energy, but it depends entirely on the availability of electrical energy produced generally through the burning of fossil fuels, not to mention the labor and energy involved in its manufacture, and the labor embodied in the costs of its purchase and maintenance. The other half of the equation is the social impact of labor-saving devices, the new social forms they enable. The time a homemaker saves by using our hypothetical electric mixer is not generally hers to use freely; rather, the device enables her to perform more tasks in the time she has, as more things are expected of her. The labor-saving devices of the modern kitchen emerged parallel with new expectations about a woman's role in the home, as a driver first in the removal of women from waged labor and, as they returned to the workforce, in the emergence of the infamous 'second shift.'

I have spent a large part of the preceding text documenting the way that FSC certification presupposes the labors of standard-making and the respective contributions (voluntary or otherwise) of environmental activists, forest workers, and marginalized communities. As a labor-saving device, the FSC logo packages those values into a single sign, easily recognizable and consumable by any party so inclined. Suchman and Cowan warn that labor-saving devices presuppose and generate new forms of human labor.

What new kind of labor does FSC certification as a labor-saving device threaten to generate, and what complexities are lost in the process?

REMATERIALIZING ETHICS

I throw all of these complications into the mix not to delegitimate the efforts of the FSC, but simply to suggest that the problem of consumer ethics or green consumption is not amenable to a technological fix. The ethical content of the project, the goodness of the good wood, remains an anthropological problem. It is not a problem to be resolved, fixed, or even answered like a question; rather, it is a problem to be tangled with, to be put in conversation with other issues in the domain of environmentalism and politics. Ong and Collier (2005) refer to the technologies in question as "global assemblages," because they pull together material and discursive artifacts across space. I'm calling it an anthropological problem as well, not just because of my disciplinary training, but because engaging with it means bringing out a particularly anthropological set of concerns regarding cultural practice, cultural difference, and knowledge.

In particular, I remain interested in the relationship between ethical trade systems and the knowledge systems that define them. Relatively few of the anthropologists and geographers dealing with various incarnations of ethical trade have much to say about the science project that is standardization (though see Guthman 2002). Science, after all, has played a pivotal part in telling us what's wrong with our natural environment (i.e. Taylor and Buttel 1992); much more complicated is the contribution of science to defining environmentally correct action and policy. What is the connection between *knowing* and *knowing-what's-right*? This may be one of those unanswerable questions. I feel a certainty, though, that one aspect of defining and engaging environmentally 'right' practice has to include resisting the impulse to constrain knowledge to accepted forms, and rejecting the boundary-policing of elite forms of knowledge. An important corollary to that imperative is, if not skepticism, at least caution towards projects that attempt to answer a visceral, material question (what is to be done about the world's forests?) with an abstraction (the FSC logo). As Barad says:

> Knowing is a direct material engagement, a practice of intra-acting with the world as part of the world in its dynamic material configuring, its ongoing articulation.... Ethics is about mattering, about taking account of the entangled materializations of which we are a part, including new configurations, new subjectivities, new possibilities—even the smallest cuts matter. (2007, 384)

Certification is a technology, and as such mobilizes expert knowledge in the most efficient manner it can. It would be foolish for me to reject the expertise of those who know more about forests and their challenges than I do, and there are plenty such experts. But an ethical technology is problematic when it attempts to substitute directly the reified expertise of others for a responsible engagement with the material issues. And yet it is precisely the reification of expertise and the passive, spatial deferral of environmental responsibility that make forest certification's technologies of value the ironic labor-saving devices that they are. Ethics and responsibility as spatial issues remain an ongoing problem for geographers (Massey 2004; Silk 1998; McEwan and Goodman 2010; Rocheleau and Roth 2007): How do our commitments travel across space, do we shoulder greater obligations to those whom we encounter face-to-face, can we imagine ethical or moral frameworks sophisticated enough to cope with the changing requirements of emerging global (or nano?) imbroglios? As a technology of value, forest certification was crafted precisely to provide an answer to the dilemma of managing care across distance. I hope that the contents of this book demonstrate, though, that the technology raises as many questions as it answers; for each new connection made and commitment sustained, a new veil of commodification is woven.

She doesn't use the term as such, but I read Barad's comments on 'mattering' as a call for a kind of cosmopolitics. Stengers's framework, and especially de la Cadena's contextualization of it as I discuss in Chapter 3, is every bit as political as it is epistemological, and provides space for an embodied politics of knowledge, perhaps even an embodied knowledge of politics. Haraway's "nourishing indigestion" (2007) speaks to the friction of cosmopolitics, the often uncomfortable but potentially transformative encounters across difference. Forest certification, for all its efforts at transparency, evades that friction. To the extent that the evasion succeeds, FSC-Chile's stakeholder process of negotiation is at risk of losing its cosmopolitical potential. Putting the FSC logo and its market power to work in different ways, the individual actors within FSC-Chile push back against that institutional evasion. The institutions of neoliberal multiculturalism in contemporary Chile (Richards 2010) trade the cosmopolitical for the cosmopolitan: Explicitly inclusive of otherness, within an implicitly predetermined frame. The basic characteristics of the cosmopolitan, according to one line of argument (i.e. Szerszynski and Urry 2006), are mobility and both the interest and ability to *consume* the other—other places, at the very least. By the same token, contemporary cosmopolitanism increasingly works with images and products from "elsewhere" rather than actual others to be encountered. Because of new communication technologies and the continuing expansion of global markets, cosmopolitanism is increasingly consumer-oriented and increasingly virtual.

Virtualism and the virtual have been recurrent themes in the cultural study of political economies (for a survey of these ideas, see Carrier and Miller 1998). Particularly salient given Barad's insistence on an ethics of "entangled materialism" is Carrier's discussion of virtual*ism* as an ideology and practice: "when people take this virtual reality to be not just a parsimonious description of what is really happening, but prescriptive of what the world ought to be; when, that is, they seek to make the world conform to their virtual vision. Virtualism, thus, operates at both the conceptual and practical levels, for it is a practical effort to make the world conform to the structures of the conceptual" (Carrier and Miller 1998, 2). West offers a relevant example in Fair Trade coffee promotional strategies, where we see distributors use disembedded depictions of ethical production and exchange in the global South to generate value "by creating virtual producers and hoping those narratives appeal to the virtual consumers that have been made for them" by the machinery of cultural marketing (West 2010). In the case of FSC-Chile, consider the forest workers and indigenous communities in the discussion of standard-setting (Chapter 3), spoken *for* but not speaking. Similarly, small forest proprietors (Chapter 5) and consumers (Chapter 2) are endlessly studied, consulted, marketed, supported, and speculated upon; this generates representations of them as collectives that then circulate throughout the process of defining and implementing certification. Certification's value depends on the traffic in these representations, even as they generate collective identity and political rationality among the actors themselves.

Speaking in terms of marketing and its subjects, especially in a restricted domain like forest certification or specialty coffee, makes capitalist virtualism seem like a relatively tame social force, but similar dynamics are at work in much larger and more fundamental ways. In the global financial crisis that began in 2008, collateralized debt obligations and other exotic financial instruments generated wealth out of nothing through the power of representation and speculation (MacKenzie 2009; LiPuma and Lee 2004). Home loans became fortunes in mortgage-backed securities through physical acts of misrepresentation: Lost mortgage notes, robo-signing, and other bits of bureaucratic friction like an evil twin to the circulating documents of forest certification's chain of custody that I discuss in Chapter 2. Although current high-profile crises illustrate the high stakes of virtual economies, Marx's essential insights into the commodity form suggest that virtual qualities of value have been central to capital since its inception. Arbitrage, after all, is the basic practice of capital accumulation, the creation of value elsewhere; certified wood likewise acquires value only in circulation to ethical consumers, a kind of arbitraging of environmental ethics. Maurer points to the abstracting technologies involved in turning the actual grains of seed-bearing grasses into a fungible commodity called "grain," an epistemic act at the heart of capital markets since the 1500s (Maurer 2005).

By the same token, Maurer is skeptical about the grounds for a "real" economy concealed behind the virtual. He argues for smaller, more precise strategies for normative engagements: "The lack of universal standards compels moral or ethical stances—themselves revisable, tentative, and often 'for now' rather than forever—toward the pragmatic resolution of specific problems of value" (2009, 263). If abstraction and the virtual are fundamental to the practices of capital across such a span of time, we should perhaps be cautious about making normative arguments for or against their newer emerging manifestations, at least on the grounds of their virtuality in and of itself. That is to say, the complexity of motivations and strategies at work in FSC-Chile are a reminder that powerful actors like the state or timber companies are not the only ones capable of putting representations to work. Critical geographies of consumption show that whatever top-down intentions certifiers may have, consumers and social movements will engage with ethical consumption programs on their own terms (Barnett et al. 2011; Clarke 2008; Littler 2008). For all that certification as market-based virtual governance implies the rationalizing and domesticating of ethical impulses and political pressures (Bauman 2007; Hartwick 2000), it also enables actors across the globe to imagine themselves interconnected and responsible to each other. Like West's romanticized virtual coffee producers, certification's global connectedness and responsibility still require political action to become manifest. In the hands of activists and companies alike, the FSC becomes another venue for the politics of pressure and imagery, or rather, a technology for bringing those pressures to bear on new publics in new spatial forms. FSC-Chile emerges in this context as neither the force of neoliberal co-optation that I inadvertently suggested to Luis Astorga (Chapter 1), nor is it necessarily the tool to transform capital from the outside as he would have had me believe. Decentering the consumer as I have tried to do here, treating certification as a strategy for generating political rationalities or relocating responsibility (Barnett et al. 2011), opens new possibilities for analysis and intervention.

FSC-Chile, for all of its tensions and contradictions as a piece of environmental governance and as a technology of value, remains an unpredictable element. This is particularly true given the rapid changes taking place in the political ecology of Chile and the region as a whole: The massive earthquake of 2010, the presidency of mercurial right-winger Sebastian Piñera, the radical language in Ecuador's new constitution, the explosive growth of REDD and other carbon initiatives in the forest sector, or the seizure of the state by indigenous social movements in Bolivia (see Escobar 2010). Understanding certification's implications in a maelstrom like the environmental politics of contemporary Latin America requires attention to particularity, openness to surprise, and patience with the dynamics of struggle both tacit and overt. I would echo Stengers's cosmopolitical proposal (2010), so well articulated to indigenous and environmental politics by de la Cadena (2010), for "slowing down reason," refusing to leap to conclusions about competing realities. This means both tempering our normative judgments about unpredictable forces like certification, while also doing our own ethical labor, rather than delegating either to technologies beyond ourselves.

My final thought is to provide a juxtaposition, a possible corrective to the elisions and erasures of a technology of value. Jasanoff calls them "technologies of humility":

> We need disciplined methods to accommodate the partiality of scientific knowledge and to act under conditions of irredeemable uncertainty. Let us call these technologies of humility. These technologies compel us to reflect on the sources of ambiguity, indeterminacy and complexity. Humility instructs us to think harder about how to reframe problems so that their ethical dimensions are brought to light, which new facts to seek and when to resist asking science for clarification. Humility directs us to alleviate known causes of people's vulnerability to harm, to pay attention to the distribution of risks and benefits, and to reflect on the social factors that promote or discourage learning. . . . It is a prescription to supplement science with the analysis of those aspects of the human condition that science cannot illuminate. It is a call for policy analysts and policy-makers to re-engage with the moral foundations for acting in the face of inevitable scientific uncertainty. (Jasanoff 2007, 33)

Rather than finding ways to insulate ourselves from our lack of knowledge, to authorize others to know for us and thereby carve out some innocence for ourselves, we need to find ways to move forward in engagement despite our not-knowing, to define baselines for what is acceptable in the world regardless of what we can know ourselves. It is also known as politics, and if no-one is innocent, then everyone is responsible.

NOTES

1. As I note in Chapter 1, I follow the argument of Whatmore and Thorne (1997) about the 'fairness' of Fair Trade.
2. By which I mean pragmatic in the philosophical sense, in the terms of Dewey (1927) or Latour (2004).

Bibliography

Aagesen, David. 1998. On the Northern Fringe of the South American Temperate Forest: The History and Conservation of the Monkey-Puzzle Tree. *Environmental History* 3(1).
Adonis, M., and L. Gil. 2001. Indoor Air Pollution in a Zone of Extreme Poverty of Metropolitan Santiago, Chile'. *Indoor+Built Environment* 10(3/4): 138–146.
Agrawal, Arun. 1995. Dismantling the Divide Between Indigenous and Scientific Knowledge. *Development and Change* 26: 413–439.
———. 2002. Indigenous Knowledge and the Politics of Classification. *International Social Science Journal* 54(3).
———. 2005. *Environmentality: Technologies of Government and the Making of Subjects*. Durham: Duke University Press.
Allen, Patricia, and Martin Kovach. 2000. The Capitalist Composition of Organic: The Potential of Markets in Fulfilling the Promise Of Organic Agriculture. *Agriculture and Human Values* 17: 221–232.
Althusser, Louis. 1971. "Ideology and Ideological State Apparatuses." In *Lenin and Philosophy and Other Essays*, L. Althusser, Ed. New York: Monthly Review Press.
Angell, Alan, and Carol Graham. 1995. Can Social Sector Reform Make Adjustment Sustainable And Equitable? Lessons from Chile and Venezuela. *Journal of Latin American Studies* 27(1): 189–219.
Appadurai, Arjun. 1986. "Commodities and the Politics of Value." In *The Social Life of Things*, A. Appadurai, Ed. Cambridge: Cambridge University Press.
Araya, Jose Conejo. 2003. La Invasión De Las Plantaciones Forestales En Chile: Efectos De La actividad forestal en la población Mapuche. Observatorio Latinoamericano de Conflictos Ambientales. Electronic document, www.wrm.org.uy/paises/Chile/invasion.pdf, accessed February 5, 2008.
Armesto, Juan, Carolina Villagrán, and Mary Kalin Arroyo. 1995. *Ecología De Los Bosques Nativos De Chile*. Santiago: Editorial Universitaria.
Astorga, Luis. 2004. El Componente Social En Certificación De Empresas Forestales. *Ambiente Y Desarrollo* 20(1): 47–52.
Auld, Graeme, L. Gulbrandsen, and C. Mcdermott. 2008. Certification Schemes and the Impacts on Forests And Forestry. *Annual Review of Environment And Resources* 33: 187–211.
Aylwin, José. 1998. "Indigenous Peoples Rights in Chile: Progresses and Contradictions in a Context Of Economic Globalization." Paper presented at the Canadian Association for Latin American and Caribbean Studies (CALACS) XXVIII Congress, Simon Fraser University, Vancouver, BC, March 19–21, 1998.
Aylwin, Jose, Andrea Coñuecar, and Elicura Chihuailaf. 2004. *El Despertar Del Pueblo Mapuche*. Santiago: Ediciones LOM.

Bael, David, and Roger Sedjo. 2006. *Toward Globalization of the Forest Products Industry: Some Trends*. RFF Discussion Paper 06–35. Washington DC: Resources for the Future.
Bañados, Francisco, and Ivan Fredes. 2003. Baja Poblacional: Mapuches De Santiago Acusan al INE de 'genocidio estadistico.' *El Mercurio* March 29. Electronic document, http://diario.elmercurio.com/nacional/nacional/noticias/2003/3/29/315170.htm?id=315170
Barad, Karen. 1996. "Meeting the Universe Halfway: Realism and Social Constructivism without Contradiction." In *Feminism, Science and the Philosophy Of Science*, Lynn Hankinson Nelson and Jack Nelson, Eds. New York: Springer, 161–194.
———. 2007. *Meeting the Universe Halfway: Quantum Physics and the Entanglement of Matter and Meaning*. Durham: Duke University Press.
Barham, Elizabeth. 2002. Towards a Theory of Values-Based Labeling. *Agriculture and Human Values* 19: 349–360.
Barnes, D., K. Openshaw, K. Smith, and R. van der Plas. 1994. What Makes People Cook with Improved Biomass Stoves? A Comparative International Review of Stove Programs. World Bank Technical Paper #242. Washington DC: World Bank.
Barnett, Clive, P. Cloke, N. Clarke, and A. Malpass. 2011. *Globalizing Responsibility: The Political Rationalities of Ethical Consumption*. London: Wiley-Blackwell.
Bauman, Zygmunt. 2007. *Consuming Life*. Cambridge: Polity.
Baviskar, Amita. 2000. "Claims to Knowledge, Claims to Control: Environmental Conflict in the Great Himalaya National Park, India." In *Indigenous Environmental Knowledge and Its Transformations*, R.F. Ellen, Bicker, and Parks, Eds. Amsterdam: Harwood Academic Publishers.
Belsky, Jill. 2003. "Unmasking the Local: Gender, Community and Politics of Community- Based Rural Ecotourism in Belize." In *Contested Nature: Power, Protected Areas and the Dispossessed*, S.R. Brechin, P. West, P. Wilshusen, and C. Fortwangler, Eds. Albany, NY: SUNY Press.
Bengoa, José. 1999. *Historia De Un Conflicto: El Estado Y Los Mapuches En El Siglo XX*. Santiago: Planeta.
———. 2000a. *Historia Del Pueblo Mapuche, Siglo XIX Y XX*. Santiago: LOM Ediciones.
———. 2000b. *La Emergencia De Indígena En América Latina*. Santiago: Fondo De Cultura Economica.
Bengoa, Jose, and Sabaag. 1997. *Los Mapuches: Comunidades Y Localidades En Chile*. Santiago: Instituto Nacional De Estadísticas.
Bennett, Elizabeth. 2001. Timber Certification: Where Is the Voice of the Biologist? *Conservation Biology* 15(2): 308–310.
Bernstein, Steven, and Benjamin Cashore. 2004. "Nonstate Global Governance: Is Forest Certification a Legitimate Alternative to a Global Forest Convention?" In *Hard Choices, Soft Law*, Kirton and Trebilcock, Eds. Pp. 33–64. London: Ashgate Publishing.
Berkes, Fikret, J. Colding, and C. Folke. 2000. The Rediscovery of Traditional Ecological Knowledge as Adaptive Management. *Ecological Applications* 10(5): 1251–1262.
Berlin, Brent. 1990. *Ethnobiological Classification: Principles of Categorization of Plants and Animals in Traditional Societies*. Princeton: Princeton University Press.
Berlin, Brent. 1992. *Ethnobiological Classification: Principles of Categorization of Plants and Animals in Traditional Societies*. Princeton: Princeton University Press.
Berlin, Brent, R. Breedlove, and P. Raven. 1966. Folk Taxonomies and Biological Classification. *Science* 154: 273–275.
Berreman, Gerald. 1966. Anemic and Emetic Analyses in Social Anthropology. *American Anthropologist* 68(2): 346–354.

Blaikie, Piers, and Harold Brookfield, Eds. 1987. *Land Degradation and Society*. London: Methuen.
Blaser, Mario. 2010. *Storytelling Globalization from the Chaco and Beyond*. Durham, NC: Duke University Press.
Bluth Solari, Alejandra. 2002. *Chile, País Forestal: Una Realidad Que Se Consolida*. Santiago: Corporación Chilena De La Madera A.G.
Bohannon, Paul, and Laura Bohannon. 1968. *Tiv Economy*. Evanston, IL: Northwestern University Press.
Bowker, Geoffrey, and Susan Leigh Star. 1999. *Sorting Things Out: Classification and Its Consequences*. Cambridge: MIT Press.
Bowyer, Joseph. 2004. Changing Realities in Forest Sector Markets. *Unasylva* 55(219): 59–65.
Boyce, James. 1994. Inequality as a Cause of Environmental Degradation. *Ecological Economics* 11: 169–178.
Braun, Bruce. 2002. *The Intemperate Rainforest: Nature, Culture and Power on Canada's West Coast*. Minneapolis: University Of Minnesota Press.
Brokensha, D., D. Warren, and O. Werner. 1980. *Indigenous Knowledge Systems and Development*. Lanham, MD: University Press of America.
Brosius, J. Peter. 1997. Endangered Forest, Endangered People: Environmentalist Representations of Indigenous Knowledge. *Human Ecology* 25(1): 47–69.
———. 1999a. Green Dots, Pink Hearts: Displacing Politics from the Malaysian Rain Forest. *American Anthropologist* 101(1): 36–57.
———. 1999b. Analyses and Interventions: Anthropological Engagements with Environmentalism. *Current Anthropology* 40(3): 277–310.
———. 2001. "Local Knowledges, Global Claims: On the Significance of Indigenous Ecologies in Sarawak, East Malaysia. In *Indigenous Traditions and Ecology*, J. A.Grim, Ed. Cambridge, MA: Harvard University Press.
———. 2006. "What Counts as Local Knowledge in Global Environmental Assessments and Conventions?" In *Bridging Scales and Knowledge Systems*, H. Reid et al., Eds. Washington DC: Island Press.
Brosius, J. Peter, Charles Zerner, and Anna Tsing. 2005. *Communities and Conservation: Histories and Politics Of Community-Based Natural Resource Management*. Lanham MD: Alta Mira Press.
Brown, Cecil. 1977. Folk Botanical Life-forms: Their Universality and Growth. *American Anthropologist* 79(2): 317–342.
Brown, Nicholas, Reed Noss, David Diamond, and Mariah Myers. 2001. Conservation Biology and Forest Certification: Working Together Toward Ecological Sustainability. *Journal of Forestry* August 2001: 18–25.
Brown, Patricia Leigh. 2008. For 'EcoMoms,' Saving Earth Begins at Home. *New York Times* February 16. Electronic document, www.nytimes.com, accessed February 2008.
Brush, Stephen. 1993. Indigenous Knowledge of Biological Resources and Intellectual Property Rights: The Role of Anthropology. *American Anthropologist* 95(3): 653–686.
Bryan, Joseph. 2001. *By Reason or by Force: Territory, State Power, and Mapuche Land Rights in Southern Chile*. Unpublished MA Thesis in Geography, UCBerkeley.
Bryant, Ray, and Michael Goodman. 2004. Consuming Narratives: The Political Ecology of 'Alternative' Consumption. *Transactions of the Institute of British Geographers* 29.
Bryant, Raymond L. 1998. Power, Knowledge and Political Ecology in the Third World: A Review. *Progress in Physical Geography* 22(1): 79–94.
Burchell, G., C. Gordon, and P. Miller, Eds. 1991. *The Foucault Effect: Studies in Governmentality*. Chicago: University of Chicago Press.

Busch, Lawrence. 2000. The Moral Economy of Grades and Standards. *Journal of Rural Studies* 16: 273–283.

Busch, Lawrence, and Keiko Tanaka. 1996. Rites of Passage: Constructing Quality in a Commodity Subsector. *Science, Technology & Human Values* 21(1): 3–27.

Calbacura, Jorge. 2006. Nación Mapuche: Contrapunto Conceptual. Working Paper Series 26, Ñuke Mapförlaget. Electronic document, www.mapuche.info/mapuint/amapuint00.htm, accessed April 4, 2008.

Callon, Michel. 2001. "Four Models for the Dynamics of Science." In *Handbook of Science and Technology Studies*, S. Jasanoff et al., Eds. Pp. 29–63. London: SAGE.

Camus, Pablo. 2005. *Ambiente, Bosques y Gestion Forestal en Chile, 1541 – 2005*. Santiago: LOM Ediciones.

Camus Gayán, Pablo. 2006. *Ambiente, Bosques Y Gestión Forestal En Chile, 1541–2005*. Santiago: LOM Ediciones.

Canguilhem, Georges. 1991. *The Normal and the Pathological*. New York: Zone Books.

Caplan, Jane, and John Torpey. 2001. *Documenting Individual Identity: The Development of State Practices in the Modern World*. Princeton, NJ: Princeton University Press.

Cardoso, Fernando Henrique. 1972. Dependency and Development in Latin America. *New Left Review* 74: 83–95.

Carlton, Jim. 2004. Once Targeted by Protesters, Home Depot Plays a Green Role. *Wall Street Journal*, August 6: A1, A6.

Carney, Judith. 1996. "Converting the Wetlands, Engendering the Environment: The Intersection of Gender with Agrarian Change in Gambia." In *Liberation Ecologies*, R. Peet and M. Watts, Eds. Pp. 165–187. New York: Routledge.

Carrere, Ricardo, and Larry Lohmann. 1996. *Pulping the South: Industrial Tree Plantations and the World Paper Economy*. London: Zed Books.

Carrier, James. 1995. *Gifts and Commodities: Exchange and Western Capitalism Since 1700*. London: Routledge.

———.1997. "Introduction." In *Meanings of the Market: The Free Market in Western Culture*. J. Carrier, Ed. Pp. 1–68. London: Berg Press.

Carrier, James, and D. Miller. 1998. *Virtualism: A New Political Economy*. Oxford: Berg.

Carruthers, David. 2001. Environmental Politics in Chile: Legacies of Dictatorship and Democracy. *Third World Quarterly* 22(3): 343–358.

Casals, Vicente Costa. 1999. La Política Forestal En Chile: Una Perspectiva Histórica. *Scripta nova: Revista electrónica de geografía y ciencias sociales* 45(16). Electronic document, www.ub.es/geocrit/sn-45-16.htm, accessed February 2008.

Cashore, Benjamin. 2002. Legitimacy and the Privatization of Environmental Governance: How Non-state Market-Driven (NSMD) Governance Systems Gain Rule-Making Authority. *Governance* 15(4): 503–539.

Cashore, Benjamin, Graeme Auld, and Deanna Newsome. 2003. Forest Certification (eco-Labeling) Programs and Their Policy-Making Authority: Explaining Divergence Among North American and European Case Studies. *Forest Policy and Economics* 5: 225–247.

———. 2004. *Governing Through Markets: Forest Certification and the Emergence of Non-state Authority*. New Haven: Yale University Press.

Cashore, Benjamin, G. Auld, S. Bernstein, and C. Mcdermott. 2007. Can Non-state Governance 'Ratchet Up' Global Environmental Standards? *Review of European Community & International Environmental Law* 16(2): 158–172.

Castree, Noel. 1995. The Nature of Produced Nature: Materiality and Knowledge Construction in Marxism. *Antipode* 27: 12–28.

———. 2001. Commodity Fetishism, Geographical Imaginations and Imaginative Geographies. *Environment and Planning A* 33: 1519–1525.

Castree, Noel, and Derek Gregory, Eds. 2006. *David Harvey: A Critical Reader*. London: Blackwell.
Castro Lucic, Milka. 2005. Challenges in Chilean Intercultural Policies: Indigenous Rights and Economic Development. *PoLAR* 28(1): 112–132.
Catalán, Rodrigo. 1999. *Pueblo Mapuche, Bosque Nativo Y Plantaciones Forestales*. Temuco: Ediciones Universidad Católica.
Catalán, Rodrigo, Petra Wilken, Angelika Kandzior, David Tecklin, and Heinrich Burschel. 2005. *Bosques Y Comunidades Del Sur De Chile*. Santiago: Editorial Universitaria.
Centro De Análisis De Políticas Públicas. 2006. *Informe País: Estado Del Medio Ambiente En Chile 2005*. Santiago: Universidad De Chile Y LOM Ediciones.
Chambers, Robert, A. Pacey, and L. Thrupp. 1989. *Farmer First: Innovation and Agricultural Research*. London: Intermediate Technology Publications.
Chapin, Mac. 2004. A Challenge to Conservationists. *Worldwatch Magazine* November/December: 17–31.
Choy, Timothy. 2005. Articulated Knowledges: Environmental Forms After Universalism's Demise. *American Anthropologist* 107(1): 5–18.
Clapp, Roger Alex. 1995a. The Unnatural History of the Monterrey Pine. *Geographical Review* 85(1): 1–19.
———. 1995b. Creating Competitive Advantage: Forest Policy as Industrial Policy in Chile. *Economic Geography* 71(3): 273–296.
———. 1998a. Regions of Refuge and the Agrarian Question: Peasant Agriculture and Plantation Forestry in Chilean Araucanía. *World Development* 26(4): 571–859.
———. 1998b. Waiting for the Forest Law: Resource-Led Development and Environmental Politics in Chile. *Latin American Research Review* 33(2): 3–37.
———. 2001. Tree Farming and Forest Conservation in Chile: Do Replacement Forests Leave Any Originals Behind? *Society and Natural Resources* 14(4): 341–356.
Clarke, Nick. 2008. From Ethical Consumerism to Political Consumption. *Geography Compass* 2(6): 1870–1884.
Clifford, James, and George Marcus. 1986. *Writing Culture: The Poetics and Politics of Ethnography*. Berkeley: University Of California Press.
Collier, Stephen, Andrew Lakoff, and Paul Rabinow. 2004. Biosecurity: Towards an Anthropology of the Contemporary. *Anthropology Today* 20(5): 3–7.
Comité Nacional Pro Defensa De La Fauna Y Flora (CODEFF). 2005. *Ciudadanos De La Tierra: Construyendo Sueños, 1968–2005*. Santiago: CODEFF.
Corporación Nacional Forestal (CONAF). 1999. *Catastro Y Evaluación De Los Recursos Vegetacionales Nativos De Chile: Informe Nacional Con Variables Ambientales*. Santiago: CONAF.
Cook, Ian. 1994. "New Fruits and Vanity: Symbolic Production in the Global Food Economy." In *From Columbus To Conagra: The Globalization of Agriculture*, A. Bonanno, L. Busch, W. Friedland, L. Gouveia, and E. Mingione, Eds. Pp. 232–248. Lawrence: University of Kansas Press.
Cook, Ian, and Paul Crang. 1996. The World on a Plate: Culinary Culture, Displacement, and Geographical Knowledges. *Journal of Material Culture* 1: 131–153.
CORMA (corporación Nacional De La Madera). 2007. *Memoria Anual 2006/2007*. Santiago: CORMA.
Cowan, Ruth Schwartz. 1983. *More Work for Mother: The Ironies of Household Technologies from the Open Hearth to the Microwave*. New York: Basic Books.
Counsell, Simon, and Kim Terje Loraas. 2002. *Trading in Credibility: The Myth and Reality of the Forest Stewardship Council*. London: Rainforest Foundation UK.
Cronon, William. 1995. The Trouble with Wilderness, or, Getting Back to the Wrong Nature. In *Uncommon Ground*, W. Cronon, Ed. Pp. 69–90. New York: Norton.

Cupples, J., and Kindon, S. 2003. Far From Being 'Home Alone': The Dynamics of Accompanied Fieldwork. *Singapore Journal of Tropical Geography* 24(2): 211–228.

Darier, Eric. 1996. Environmental Governmentality: The Case of Canada's Green Plan. *Environmental Politics* 5(4): 585–606.

Dauvergne, Peter. 1997. *Shadows in the Forest: Japan and the Politics of Timber in Southeast Asia*. Cambridge: MIT Press.

Dawson, Jane. 1996. *Eco-nationalism: Anti-nuclear Activism and National Identity in Russia, Lithuania and Ukraine*. Durham, NC: Duke University Press.

De La Cadena, Marisol. 2010. Indigenous Cosmopolitics in the Andes: Conceptual Reflections Beyond "Politics." *Cultural Anthropology* 25(2): 334–370.

Dewey, John. 1927. *The Public and Its Problems*. Athens: Ohio University Press.

Dimaggio, Paul, and Walter Powell. 1983. The Iron Cage Revisited: Institutional Isomorphism and Collective Rationality in Organizational Fields. *American Sociological Review* 48: 147–160.

Dirks, Nicholas. 1992. *Colonialism and Culture*. Ann Arbor: University Of Michigan Press.

Dove, Michael. 2000. "The Life-cycle of Indigenous Knowledge, and the Case of Natural Rubber Production." In *Indigenous Environmental Knowledge and Its Transformations*, R. Ellen, P. Parkes, and A. Bicker, Eds. Pp. 213–252. Amsterdam: Harwood Academic.

———. 2002. Hybrid Histories and Indigenous Knowledge Among Asian Rubber Smallholders. *International Social Science Journal* 54(173): 349–359.

Drew, Joshua, and Adam Henne. 2006. Conservation Biology and Traditional Ecological Knowledge: Integrating Academic Disciplines for Better Conservation Practice. *Ecology and Society* 11(2): 34.

Dunn, Elizabeth. 2003. Trojan Pig: Paradoxes of Food Safety Regulation. *Environment and Planning A* 35: 1493–1511.

———. 2005. "Standards and Person-Making in East Central Europe." In *Global Assemblages: Technology, Politics and Ethics as Anthropological Problems*, A. Ong and S. Collier, Eds. Pp. 173–193. Malden: Blackwell Publishing.

Duran, Teresa, E. Parada, and N. Carrasco. 2000. *Acercamientos Metodológicos Hacia Pueblos Indígenas. Una Experiencia Reflexionada Desde La Araucanía*. Temuco: Universidad Católica De Temuco Y Editorial LOM.

Durston, John, and Daniel Duhart. 2003. *Formación Y Perdida De Capital Social Comunitario Mapuche: Cultura, Clientalismo, Y Empoderamiento En Dos Comunidades, 1999–2002*. Port-of-Spain: UN Economic Commission for Latin America and the Caribbean.

Dutfield, Graham. 2002. *Intellectual Property Rights, Trade, and Biodiversity*. London: Earthscan Limited.

Economist. 2006. Emerging Economies. January 19. Electronic document, www.economist.com/opinion/displaystory.cfm?story id=5411977, accessed April 7, 2008.

Ellen, Roy, and H. Harris. 2000. "Introduction." In *Indigenous Environmental Knowledge and Its Transformations*, R.F. Ellen, P. Parkes, and A. Bicker, Eds. Amsterdam: Harwood.

Elliot, Chris, and Rodolphe Schlaepfer. 2001. Understanding Forest Certification Using the Advocacy Coalition Framework. *Forest Policy and Economics* 2: 257–266.

Ellis, David, and Paige West. 2004. "Local History as Indigenous Knowledge: Aeroplanes, Conservation and Development in Haia and Maimafu, Papua New Guinea." In *Investigating Local Knowledge: New Directions, New Approaches*, A. Bicker, P. Sillitoe, and J. Pottier, Eds. Pp. 105–127. London: Ashgate.

Escobar, Arturo. 1995. *Encountering Development: The Making and Unmaking of the Third World*. Princeton: Princeton University Press.

———. 2001. Whose Knowledge, Whose Nature? Biodiversity, Conservation, and the Political Ecology of Social Movements. *Journal of Political Ecology* 5: 53–82.
———. 2008. *Territories of Difference: Place, Movements, Life, Redes*. Durham, NC: Duke University Press.
———. 2010. Latin America at a Crossroads. *Cultural Studies* 24(1): 1–65.
Evans-schultes, Richard. 1995. *Ethnobotany: Evolution of a Discipline*. Portland: Timber Press.
Ezzati, Majid. 2005. Indoor Air Pollution and Health in Developing Countries. *Lancet* 366: 105–106.
Ezzati, Majid, and Daniel M Kammen. 2002. Household Energy, Indoor Air Pollution, and Health in Developing Countries: Knowledge Base for Effective Interventions. *Annual Review of Energy and the Environment* 27: 233–270.
Fairhead, James, and Melissa Leach. 1995. False Forest History, Complicit Social Analysis: Rethinking Some West African Environmental Narratives. *World Development* 23(6): 1023–1035.
FAO (Food and Agriculture Organization). 2007. *State of the World's Forests 2007*. Rome: FAO.
Faron, Louis. 1986 [1940]. *The Mapuche Indians of Chile*. Long Grove IL: Waveland Press.
Faubion, James. 2001. Toward an Anthropology of Ethics: Foucault and the Pedagogies of Autopoiesis. *Representations* 74 (Spring): 83–104.
Federovisky, Sergio. 2007. *El Medio Ambiente No Le Importa A Nadie*. Buenos Aires: Editorial Planeta.
Ferguson, James. 1994. *The Anti-politics Machine: Development, Depoliticalization and Bureaucratic Power in Lesotho*. Minneapolis: University of Minnesota Press.
Ford, Richard. 1999. "Ethnoecology Serving the Community." In *Ethnoecology: Situated Knowledge/Located Lives*, V. Nazarea, Ed. Tucson: University of Arizona Press.
Forest Stewardship Council. 1994. Forest Stewardship Council A.C. Bylaws. FSC Secretariat general publication, available at www.fsc.org. Accessed November 2005.
Forest Stewardship Council (FSC). 2005. Leading Our World Toward Responsible Forest Stewardship: A Progress Report. Forest Stewardship Council. Bonn.
Forest Stewardship Council—US. 2006. *FSC-US Prospectus 2006*. Electronic document, www.fscus.org/about_us/prospectus_2006.php, accessed February 2008.
Fortun, Kim. 2001. *Advocacy After Bhopal: Environmentalism, Disaster, New World Orders*. Chicago: University of Chicago Press.
Foucault, Michel. 1978. *The History of Sexuality, Volume 1: The Will to Knowledge*. London: Penguin.
———. 1980. *Power/Knowledge: Selected Interviews 1972–1977*. New York: Pantheon Books.
———. 1982. The Subject and Power. *Critical Inquiry* 8(4): 777–795.
———. 1991. "Governmentality." In *The Foucault Effect*, G. Burchell, C. Gordon, aAnd P. Miller, Eds. Pp. 87–104. Chicago: University of Chicago Press.
Frank, Andre Gunter. 1975. *On Capitalist Underdevelopment*. Bombay: Oxford University Press.
Frias, Gisela. 2003. *Invasión Forestal: Khla Nagnegai Taiñ Weichangepan*. Toronto: IDRC.
Friedman, Jonathan. 2004. "Globalization." In *Companion to the Anthropology of Politics*, D. Nugent and J. Vincent, Eds. Pp. 179–197. Oxford: Blackwell.
Gabrielson, Teena, and Kaitlyn Parady. 2010. Corporeal Citizenship: Rethinking Green Citizenship through the Body. *Environmental Politics* 19(3): 374–392.
Gadgil Madhav. 2000. New Meanings for Old Knowledge: The People's Biodiversity Register Program. *Ecological Applications* 10(5): 1307–1317.

Gieryn, Thomas. 1999. *Cultural Boundaries of Science: Credibility on the Line.* Chicago: University of Chicago Press.

Godelier, Maurice. 1977. *Perspectives in Marxist Anthropology.* Cambridge: Cambridge University Press.

Gonzalez, Gustavo. 2002. Isabel Allende Helps Defend Native Forests. *Environment Chile,* July 24. Electronic document, www.ips.org, accessed November 10, 2010.

Goodman, Michael. 2004. Reading Fair Trade: Political Ecological Imaginary and the Moral Economy of Fair Trade Foods. *Political Geography* 23: 891–915.

Gordillo, Gaston. 2006. The Crucible of Citizenship: ID-Paper Fetishism in the Argentinian Chaco. *American Ethnologist* 33(2): 162–175.

Gottlieb, Alma, Philip Graham, and Nathaniel Gottlieb-Graham. 1998. Infants, Ancestors and the Afterlife: Fieldwork's Family Values. *Anthropology and Humanism* 23(2): 121–126.

Grace, V., and L. Arnoux. 1998. Clean-Burning Fuel for Use in Woodstoves: Feminist Politics, Community Development and Global Sustainability. *Community Development Journal* 33: 260–269.

Graeber, David. 2001. *Toward an Anthropological Theory of Value.* New York: Palgrave.

———. 2004. *Fragments of an Anarchist Anthropology.* Chicago: Prickly Paradigm Press.

Gregory, Robin. 2000. Using Stakeholder Values to Make Smarter Environmental Management Decisions. *Environment* 42: 34–44.

Gullison, R.E. 2003. Does Forest Certification Conserve Biodiversity? *Oryx* 37(2): 153–165.

Gupta, Akhil. 1998. *Postcolonial Developments: Agriculture in the Making of Modern India.* Durham: Duke University Press.

Gupta, Akhil, and James Ferguson. 1992. Beyond 'Culture': Space, Identity, and the Politics of Difference. *Cultural Anthropology* 7(1): 6–23.

Guthman, Julie. 1998. Regulating Meaning, Appropriating Nature: The Codification of California Organic Agriculture. *Antipode* 30(2): 135–154.

———. 2000. Raising Organic: An Agro-ecological Assessment of Grower Practices in California. *Agriculture and Human Values* 17: 257–266.

———.2002. Commodified Meanings, Meaningful Commodities: Re-thinking Production Consumption Links through the Organic System of Provision. *Sociologia Ruralis* 42(4): 295–311.

———. 2004. Back to the Land: The Paradox of Organic Food Standards. *Environment and Planning A* 36: 511–528.

Haas, Peter, Robert Keohane, and Marc Levy. 1993. *Institutions for the Earth: Sources of Effective Environmental Protection.* Cambridge: MIT Press.

Haener, M.K., and M.K. Luckert. 1998. Forest Certification: Economic Issues and Welfare Implications. *Canadian Public Policy* 24(2): S83–S94.

Hagerty, James. 1999. Home Depot Vows to Change Policy on "Sensitive" Wood Use. *Wall Street Journal,* August 27.

Hale, Charles. 2006. Activist Research Vs. Cultural Critique: Indigenous Land Rights and the Contradictions of Politically Engaged Anthropology. *Cultural Anthropology* 21(1): 96– 120.

Haraway, Donna Jeanne. 1989. Primate Visions: Gender, Race, and Nature in the World of Modern Science. Psychology Press.

———. 1991. *Simians, Cyborgs, and Women.* New York: Routledge.

———. 1992. "The Promises of Monsters: A Regenerative Politics for Inappropriate/d Others." In *Cultural Studies,* L. Grossberg, C. Nelson, and P. Treichler, Eds. Pp. 295–337. New York: Routledge.

———. 1997. "Mice into Wormholes: A Comment on the Nature of No Nature." In *Cyborgs and Citadels: Anthropological Interventions in Emerging Sciences and Technologies.* Santa Fe: School of American Research.

———. 2008. *When Species Meet*. Minneapolis: University Of Minnesota Press.
Hardt, Michael, and Antonio Negri. 2000. *Empire*. Cambridge: Harvard University Press.
Harper, Krista. 2005. "Wild Capitalism" and "Ecocolonialism": A Tale of Two Rivers. *American Anthropologist* 107(2): 221–233.
Hartwig, Fernando. 1994. *La Tierra Que Recuperamos*. Santiago: Editorial Los Andes.
Hartwick, Elaine R. 2000. Towards a Geographical Politics of Consumption. *Environment and Planning A* 32(7): 1177–1192.
Harvey, David. 1990. Between Space and Time: Reflections on the Geographical Imagination. *Annals of the Association of American Geographers* 80: 418–434.
———. 1996. *Justice, Nature, and the Geography Of Difference*. Cambridge: Blackwell.
Hawken, Paul, Amory Lovins, and L. Hunter Lovins. 2000. *Natural Capitalism: Creating the Next Industrial Revolution*. San Francisco: Back Bay Books.
Hayden, Cori. 2003. *When Nature Goes Public: The Making and Unmaking of Bioprospecting in Mexico*. Princeton: Princeton University Press.
Hayes-Conroy, A. and J. Hayes-Conroy. 2008. Taking Back Taste: Feminism, Food, and Visceral Politics. *Gender, Place and Culture* 15(5): 461–473.
Hayes-Conroy, J. 2009. Get Control of Yourselves! The Body as ObamaNation'. *Environment and Planning A* 41: 1020–1025.
Hayward, Steven. 1998. The Shocking Success of Welfare Reform. *Policy Review* 87. Electronic document www.hoover.org/publications/policyreview/3566722.html, accessed February 2008.
Helmreich, Stefan, and Eben Kirksey. 2010. The Emergence of Multispecies Ethnography. *Cultural Anthropology* 25(4): 545–576.
Henne, Adam, and Teena Gabrielson. 2012. "Chile Is Timber Country: Citizenship, Justice and Scale in the Chilean Native Forest Market Campaign." In *Environment and Citizenship in Latin America*, Hannah Witman and Alex Latta, Eds. New York: CEDLAP/Berghahn Books.
Herod, Andrew, and Melissa Wright. 2002. *Geographies of Power: Placing Scale*. Oxford: Blackwell.
Hess, David J., Linda L. Layne, and Arie Rip, Eds. 1992. Knowledge and Society: The Anthropology of Science and Technology. Vol. 9. JAI Press.
Heyman, Josiah. 2004. Ports of Entry as Nodes in the World System. *Identities* 11(3): 303–327.
Hobart, Mark. 1993. "Introduction: The Growth of Ignorance?" In *An Anthropological Critique of Development*, M. Hobart, Ed. London: Routledge.
Hoffman, Adriana, Ed. 1998. *La Tragedia Del Bosque Chileno*. Santiago: Ocho Libros Editores.
Hughes, Alex. 2000. Retailers, Knowledges, and Changing Commodity Networks: The Case of the Cut Flower Trade. *Geoforum* 31: 175–190.
Hughes, Alex, and S. Reimer, Eds. 2004. *Geographies of Commodity Chains*. London: Routledge.
Hunn, Eugene. 1982. The Utilitarian Factor in Folk Botanical Classification. *American Anthropologist* 84(4): 830–847.
ICEFI (initiava Chilena Para Certification Forestal Independiente). 2002. Boletin Electronico #16, October-November.
Igoe, James. 2002. *Conservation and Globalization: A Study of National Parks and Indigenous Communities from East Africa to South Dakota*. Belmont, CA: Wadsworth Publishing.
INFOR (instituto Forestal). 2006. *El Sector Forestal Chileno 2006*. Santiago: INFOR. Electronic document, www.infor.cl, accessed April 4, 2008.
———. 2007. *Mercado Forestal*. #30, October 2007. INFOR and Ministerio De Agricultura. Electronic document, www.infor.cl, accessed April 4, 2008.

Ingold, Timothy. 1993. "Globes and Spheres: The Topology of Environmentalism." In *Environmentalism: The View from Anthropology*, Kay Milton, Ed. Pp. 31–42. London: Routledge.

———. 2000. *The Perception of the Environment: Essays on Livelihood, Dwelling and Skill*. London: Routledge.

———. 2004. Culture on the Ground: The World Perceived through the Feet. *Journal of Material Culture* 9(3): 315–340.

Irwin, Alan. 2006. The Politics of Talk: Coming to Terms with the 'New' Scientific Governance. *Social Studies of Science* 36(2): 299–320.

Isbell, Billie Jean. 1985. *To Defend Ourselves: Ritual and Ecology in an Andean Village*. Prospect Heights, IL: Waveland Press.

Jackson, Peter. 1999. Commodity Cultures: The Traffic in Things. *Transactions of the Institute of British Geographers* 24: 95–108.

James, A., J. Hockey, and A. Dawson, Eds. 1997. *After Writing Culture: Epistemology and Praxis in Contemporary Anthropology*. London and New York: Routledge.

Jameson, Fredric. 2002. *A Singular Modernity: Essays on the Ontology of the Present*. London: Verso.

Jasanoff, Sheila. 1987. Contested Boundaries in Policy-Relevant Science. *Social Studies of Science* 17: 195–230.

———. 1990. *The Fifth Branch: Science Advisors as Policymakers*. Cambridge: Cambridge University Press.

———. 2003. Technologies of Humility: Citizen Participation in Governing Science. *Minerva* 41: 223–244.

———. 2007. Technologies of Humility. *Nature* 450 (November 1): 33.

Joss, Simon, and Sergio Berlucci. 2002. *Participatory Technology Assessment: European Perspectives*. London: Center for the Study of Democracy.

Kelty, Chris. 2005. Geeks, Social Imaginaries, and Recursive Publics. *Cultural Anthropology* 20(2): 185–214.

Kindon, Sara, Rachel Pain, and Mike Kesby, Eds. 2007. *Participatory Action Research Approaches and Methods: Connecting People, Participation and Place*. Routledge.

Kirton, John, and Michael Trebilcock. 2004. *Hard Choices, Soft Law: Voluntary Standards in Global Trade, Environment and Social Governance*. London: Ashgate Publishing.

Klein, Naomi. 2009. *No Logo*. New York: Macmillan.

Klubock, Timothy M. 2004. "Labor, Land, and Environmental Change in the Forestry Sector in Chile, 1973–1998." In *Victims of the Chilean Miracle*, P. Winn, Ed. Durham: Duke University Press.

———. 2006. The Politics of Forests and Forestry on Chile's Southern Frontier, 1880s–1940s. *Hispanic American Historical Review* 86(3): 535–570.

Kohn, Eduardo. 2007. How Dogs Dream: Amazonian Natures and the Politics of Transspecies Engagement. *American Ethnologist* 31(4): 3–24.

Kun, Zhang, Lu Wenming, and Osamu Hashiramoto. 2006. *Demand and Supply of Wood Products in China*. FAO Forest Products Working Paper 1. Rome: Food and Agriculture Organization.

Kurtz, Hilda. 2002. "The Politics of Environmental Justice as a Politics of Scale." In *Geographies of Power*, A. Herod and M. Wright, Eds. Pp. 249–273. Oxford: Blackwell.

———. 2003. Scale Frames and Counter Scale Frames: Constructing the Social Grievance of Environmental Injustice. *Political Geography* 22: 887–916.

Ladio, Ana, and Mariana Lozada. 2000. Edible Wild Plant Use in a Mapuche Community of Northwestern Patagonia. *Human Ecology* 28(1): 53–71.

Ladio, Ana H., and Mariana Lozada. 2004. Patterns of Use and Knowledge of Wild Edible Plants in Distinct Ecological Environments: A Case Study of a Mapuche

Community From Northwestern Patagonia. *Biodiversity & Conservation* 13(6): 1153–1173.
Lakoff, Andrew, and Stephen Collier. 2004. Ethics and the Anthropology of Modern Reason. *Anthropological Theory* 4(4): 419–434.
Lara, Antonio, D. Soto, J. Armesto, P. Donoso, C. Wernli, L. Nahuelhual, and F. Squeo, Eds. 2003. *Componentes Científicos Clave Para Un Política Nacional Sobre Usos, Servicios Y Conservación De Los Bosques Nativos Chilenos*. Valdivia: Universidad Austral De Chile Y Iniciativa Científica Milenio De Mideplan.
Latour, Bruno. 1987. *Science in Action: How to Follow Scientists and Engineers through Society*. London: Open University Press.
———. 1993. *We Have Never Been Modern*. Cambridge: Harvard University Press.
———. 2004. *Politics of Nature: How to Bring the Sciences into Democracy*. Cambridge: Harvard University Press.
———. 2005. *Reassembling the Social*. New York: Oxford University Press.
———. 2010. *On the Modern Cult of the Factish Gods*. Durham: Duke University Press.
Law, John, and Annemarie Mol. 2001. Situating Technoscience: An Inquiry into Spatialities. *Environment and Planning D: Society And Space* 19: 609–621.
Li, Tania. 2000. "Locating Indigenous Environmental Knowledge in Indonesia." In *Indigenous Environmental Knowledge and Its Transformations*, R. Ellen, P. Parkes, and A. Bicker, Eds. Pp. 212–250. Amsterdam: Harwood Academic.
Lipschutz, Ronnie. 2001. Why Is There No International Forestry Law? An Examination of International Forestry Regulation, Public And Private. *UCLA Journal of Environmental Law And Policy* 19:153–180.
Lipuma, Edward, and Benjamin Lee. 2004. *Financial Derivatives and the Globalization of Risk*. Durham: Duke University Press.
Littler, Jo. 2008. *Radical Consumption: Shopping For Change in Contemporary Culture: Shopping for Change in Contemporary Culture*. McGraw-Hill International.
Lockie, Stewart. 2002. The Invisible Mouth: Mobilizing the Consumer in Food Production-Consumption Networks. *Sociologia Ruralis* 42(4): 278–294.
Loveman, Brian. 2001. *Chile: The Legacy of Hispanic Capitalism*. New York and London: Oxford University Press.
Lowe, Celia. 2005. *Wild Profusion: Biodiversity Conservation in an Indonesian Archipelago*. Princeton: Princeton University Press.
Lowe, Celia. 2006. *The Wild Profusion: Biodiversity Conservation in an Indonesian Archipelago*. Princeton, NJ: Princeton University Press.
Luke, Timothy. 1999. On Environmentality: Geo-power and Eco-knowledge in the Discourses of Contemporary Environmentalism. *Cultural Critique* 31: 57–81.
Lutter, Randall, and Jason Shogren. 2004. *Painting the White House Green: Rationalizing Environmental Policy Inside the Executive Office of the President*. Baltimore: Johns Hopkins University Press.
Lutz, Catherine, and Jane Collins. 1993. *Reading National Geographic*. Chicago: University of Chicago Press.
Mackenzie, Donald. 2009. *Material Markets: How Economic Agents Are Constructed*. New York: Oxford University Press.
Mackenzie, Donald, and Judy Wajcman, Eds. 1999. *The Social Shaping of Technology, 2nd Edition*. New York: McGraw.
Macqueen, Duncan, Annie Dufey, and Bindi Patel. 2006. *Exploring Fair Trade Timber: A Review of Issues in Current Practice, Institutional Structures and Ways Forward*. IIED Small and Medium Forestry Enterprise Series No. 19. Edinburgh: IIED.
Maddox, Richard. 1997. "Bombs, Bikinis, and the Popes of Rock'n'roll: Reflections On Resistance, the Play of Subordinations, and Liberalism in Andalusia and Academia, 1983–1995." In *Culture, Power, Place: Explorations in Critical*

Anthropology, A. Gupta and J. Ferguson, Eds. Pp. 277–290. Durham: Duke University Press.

Mallon, Florencia. 2005. *Courage Tastes of Blood: The Mapuche Community of Nicolás Ailío and the Chilean State, 1906–2001*. Durham: Duke University Press.

Marimán, Pablo. 1997. *Pueblos Indígenas: Educación Y Desarrollo*. Instituto De Estudias Indígenas, Universidad De La Frontera: Temuco.

Marimán, Pablo, Sergio Caniuqueo, José Millalén, and Rodrigo Levil. 2006. *!Escucha, Winka! Cuatro Ensayos De Historia Mapuche Y Un Epílogo Sobre El Futuro*. Santiago: Ediciones LOM.

Marx, Karl. 1967. *Capital: A Critique of Political Economy*. New York: International Publishers.

Maser, Chris, and Walter Smith. 2001. *Forest Certification in Sustainable Development: Healing the Landscape*. Boca Raton: Lewis Publishers.

Massey, Doreen. 2004. Geographies of responsibility. *Geografiska Annaler: Series B, Human Geography* 86(1): 5–18.

Maurer, Bill. 2005. *Mutual Life, Ltd.: Islamic Banking, Alternative Currencies, Lateral Reason*. Princeton: Princeton University Press.

Maybury-Lewis, David. 1992. *Millennium: Tribal Wisdom and the Modern World*. New York: Viking Penguin.

Mcafee, K. 1999. Selling Nature to Save It? Biodiversity and Green Developmentalism. *Environment and Planning D* 17(2): 384–397.

Mccarthy, James. 2005. Scale, Sovereignty, and Strategy in Environmental Governance. *Antipode* 37(4): 731–753.

Mccarthy, James, and Scott Prudham. 2004. Neoliberal Nature and the Nature of Neoliberalism. *Geoforum* 35: 275–283.

McDermott, Constance L., Benjamin Cashore, and Peter Kanowski. 2009. Setting the Bar: An International Comparison of Public and Private Forest Policy Specifications and Implications for Explaining Policy Trends. *Environmental Sciences* 6(3): 217–237.

McEwan, Cheryl, and Michael K. Goodman. 2010. Place Geography and the Ethics of Care: Introductory Remarks on the Geographies of Ethics, Responsibility and Care. *Ethics, Place and Environment* 13(2): 103–112.

Meidinger, Errol. 2001. Environmental Certification Programs and US Environmental Law: Closer than You May Think. *Environmental Law Reporter* 31: 10162–10179.

———. 2003. "Forest Certification as a Global Civil Society Regulatory Institution." In *Social and Political Dimensions of Forest Certification*, E. Meidinger, C. Elliot, and G. Oesten, Eds. SUNY Buffalo Legal Studies Research Paper #2015-007. Buffalo, NY.

Meidinger, Errol, Chris Elliot, and Gerhard Oesten. 2003. "The Fundamentals of Forest Certification." In *Social and Political Dimensions of Forest Certification*, E. Meidinger, C. Elliot, and G. Oesten, Eds. SUNY Buffalo Legal Studies Research Paper #2015-007. Buffalo, NY: pp. 3–26.

Mendes, Chico, and Tony Gross. 1989. *Fight for the Forest: Chico Mendes in His Own Words*. London: Latin American Bureau.

Mehta, S., and C. Shahpar. 2004. The Health Benefits of Interventions to Reduce Indoor Pollution from Solid Fuel Use: A Cost-Effectiveness Analysis. *Energy for Sustainable Development* 8(3): 53–59.

Miller, Daniel. 1995a. *Acknowledging Consumption*. London: Routledge.

———.1995b. Consumption and Commodities. *Annual Review of Anthropology* 24: 141–161.

———. 2003. Could the Internet Defetishize the Commodity? *Environment and Planning D* 21: 359–372.

Miller, Peter, and Nikolas Rose. 2008. *Governing the Present: Administering Economic, Social and Personal Life.* London: Polity.

Mintz, Sidney. 1985. *Sweetness and Power: The Place of Sugar in Modern History.* New York: Viking Press.

Mohr, Lois, Deborah Webb, and Katherine Harris. 2001. Do Consumers Expect Companies to Be Socially Responsible? The Impact of Corporate Social Responsibility on Buying Behavior. *Journal of Consumer Affairs* 35(1): 45–72.

Moock, Joyce, and Robert Rhoades. 1992. *Diversity, Farmer Knowledge, and Sustainability.* Ithaca: Cornell University Press.

Moore, Donald. 1996. "Marxism, Culture, and Political Ecology." In *Liberation Ecologies*, R. Peets and M. Watts, Eds. Pp. 125–147. London: Routledge.

Morales, Roberto Eduardo. 2000. "Los Mapuche Escritos Por Antropólogos Chilenos: Un Caso De Autorías Anónimas." In *Actas Del Tercer Congreso Chileno De Antropología*. Tomo I: 297–305. Santiago: Colegio De Antropólogos De Chile.

Moran, Katy, Steven King, and Thomas Carlson. 2001. Biodiversity Prospecting: Lessons and Prospects. *Annual Review of Anthropology* 30: 505–526.

Mutersbaugh, Tad. 2002. The Number Is the Beast: A Political Economy of Organic Coffee Certification and Producer Unionism. *Environment and Planning A* 34: 1165–1184.

Nabhan, Gary Paul. 2000. Interspecific Relations Affecting Endangered Species Recognized by O'Odham and Comcaac Cultures. *Ecological Applications* 10(5): 1288–1295.

Nadasdy, Paul. 1999. The Politics of TEK: Power and the 'Integration' of Knowledge. *Arctic Anthropology* 36(1–2): 1–18.

———. 2003. *Hunters and Bureaucrats: Power, Knowledge, and Aboriginal-State Relationships in the Southwest Yukon.* Vancouver: University Of British Columbia Press.

———. 2007. The Gift in the Animal: The Ontology of Hunting and Human-Animal Sociality. *American Ethnologist* 31(4): 25–43.

Nancy, Jean-luc. 2001. The Two Secrets of the Fetish. *Diacritics* 31(2): 2–8.

Naples, Nancy. 2003. *Feminism and Method: Ethnography, Discourse Analysis, and Activist Research.* New York: Routledge.

Nebel, Gustav, Lincoln Quevedo, Jette Bredahl Jacobsen, and Finn Helles. 2005. Development and Economic Significance of Forest Certification: The Case of FSC in Bolivia. *Forest Policy and Economics* 7: 175–186.

Neira, Eduardo, Hernán Verscheure, and Carmen Revenga. 2002. *Chile's Frontier Forests: Conserving a Global Treasure.* Global Forest Watch, World Resources Institute.

Newsom, Deanna, Volker Bahn, and Benjamin Cashore. 2005. Does Forest Certification Matter? An Analysis of Operation-Level Changes Required During the SmartWood Certification Process in the United States. *Forest Policy and Economics* 9(3): 197–208.

Nigh, Ronald. 1997. Organic Agriculture and Globalization: A Maya Associative Corporation in Chiapas, Mexico. *Human Organization* 56: 427–436.

Nygren, Anna. 1999. Local Knowledge in the Environment-Development Discourse, from Dichotomies to Situated Knowledges. *Critiques of Anthropology* 19(3): 267–288.

Oates, John. 1999. *Myth and Reality in the Rainforest.* Berkeley: University of California.

O'donnell, Jayne, and Christine Dugas. 2007. More Retailers Go for Green—The Eco Kind. *USAToday* April 17.

OECD. 2005. OECD Environmental Performance Reviews: Chile. Paris: OECD.

Ong, Aiwa, and Stephen Collier, Eds. 2005. *Global Assemblages: Technology, Politics and Ethics as Anthropological Problems.* Malden: Blackwell Publishing.

Otero, Luis. 1999. *Manual De Manejo Sustentable De Bosques Nativos Para Pequeños Proprietarios*. Santiago: Comité Nacional Pro Defensa De La Fauna Y Flora.
———. 2006. *La Huella Del Fuego: Historia De Los Bosques Nativos, Poblamiento Y Cambios En El Paisaje Del Sur De Chile*. Santiago: Pehuen.
Ozinga, Saskia. 2001. *Behind the Logo: An Environmental and Social Assessment of Forest Certification Schemes*. Moreton-in-Marsh, UK: FERN.
Paglia, Todd. 2007. "The New Environmentalism: Using Corporate Power for Social Change." Presentation to the Commonwealth Club of California, March 29. Electronic audio recording, www.commonwealthclub.org/archive/?monthRecording= Mar&yearRecording#, accessed February 2008.
Paley, Julia. 2001. *Marketing Democracy: Power and Social Movements in Post-dictatorship Chile*. Berkeley: University Of California Press.
Parsons, Talcott. 1937. *The Structure of Social Action*. New York: Free Press.
Paulson, Susan, Lisa Gezon, and Michael Watts, Eds. 2005. *Political Ecology Across Spaces, Scales, and Social Groups*. New Brunswick: Rutgers University Press.
Peluso, Nancy. 1994. *Rich Forests, Poor People: Resource Control and Resistance in Java*. Berkeley: University of California Press.
Peters, Pauline. 1996. 'Who's Local Here? The Politics of Participation in Development. *Cultural Survival Quarterly* 20(3): 22–60.
Pierotti, Raymond, and Daniel Wildcat. 2000. Traditional Ecological Knowledge: The Third Alternative. *Ecological Applications* 10(5): 1333–1340.
Pigg, Stacy Leigh. 1992. Constructing Social Categories through Place: Social Representations and Development in Nepal. *Comparative Studies in Society and History* 34: 491–513.
Pinto Rodriguez, Jorge. 2003. *La Formación Del Estado Y La Nación, Y El Pueblo Mapuche: De La Inclusión A La Exclusión*. Santiago: DIBAM.
Poncelet, Eric. 2004. *Partnering for the Environment: Multistakeholder Collaboration in a Changing World*. Lanham, MD: Rowman and Littlefield.
Poore, D. P. 2003. *Changing Landscapes: The Development of the International Tropical Timber Organization and Its Influence on Tropical Forest Management*. Earthscan, London, UK.
Posey, Darrell Addison. 1985. Indigenous Management of Tropical Forest Ecosystems: The Case of the Kayapo Indians of the Brazilian Amazon. *Agroforestry Systems* 3(2): 139–158.
Power, Michael. 1997. *The Audit Society: Rituals of Verification*. Oxford: Oxford University Press.
Rabinow, Paul. 2003. *Anthropos Today*. Princeton: Princeton University Press.
Rabinow, Paul, and Nikolas Rose. 2006. Biopower Today. *Biosocieties* 1: 195–217.
Raffles, Hugh. 2002a. Intimate Knowledge. *International Social Science Journal* 54(173): 325–335.
———. 2002b. *In Amazonia: A Natural History*. Princeton: Princeton University Press.
Raison, R. John, Alan G. Brown, and David W. Flinn. 2001. *Criteria and Indicators for Sustainable Forest Management*. New York: CABI Publishing And IUFRO.
Rametsteiner, Ewald. 2002. The Role of Governments in Forest Certification—A Normative Analysis Based on New Institutional Economics Theories. *Forest Policy and Economics* 4: 163–173.
Rametsteiner, Ewald, and M. Simula. 2003. Forest Certification—An Instrument to Promote Sustainable Forest Management? *Journal of Environmental Management* 67: 87–98.
Rappaport, Roy. 1968. *Pigs for the Ancestors: Ritual in the Ecology of a Papua New Guinea People*. New Haven: Yale University Press.
———. 1993. Distinguished Lecture in General Anthropology: The Anthropology of Trouble. *American Anthropologist* 95(2): 295–303.

Raynolds, Laura. 2000. Re-embedding Global Agriculture: International Organic and Fair Trade Movements. *Agriculture and Human Values* 17: 297–309.

———. 2002. Consumer/Producer Links in Fair Trade Coffee Networks. *Sociologia Ruralis* 42(4): 404–424.

Red Nacional De Acción Ecológica (RENACE). 2002. *Bosque Sustentable: Marco De Referencia Y Propuestas Para La Formulación De Un Ley De Conservación Y Manejo Sustentable Del Bosque.* Santiago: RENACE Y LOM Ediciones.

Redford, Kent, and A. Stearman. 1993. Forest-Dwelling Native Amazonians and the Conservation of Biodiversity: Interests in Common or in Collision? *Conservation Biology* 7(2): 248–255.

Rehfuess, E., S. Mehta, and A. Pruss-Ustun. 2006. Assessing Household Solid Fuel Use: Multiple Implications for the Millenium Development Goals. *Environmental Health Perspectives* 114: 373–378.

Reid, Walter, Fikret Berkes, Thomas Wilbanks, and Doris Capistrano. 2006. *Bridging Scales and Knowledge Systems: Concepts and Applications in Ecosystem Assessment.* Washington: Island Press.

Rice, Robert A. 2001. Noble Goals and Challenging Terrain: Organic and Fair Trade Coffee Movements in the Global Marketplace. *Journal of Agricultural and Environmental Ethics* 14(1): 39–66.

Richards, Patricia. 2006. The Politics of Difference and Women's Rights: Lessons from Pobladoras and Mapuche Women in Chile. *Social Politics* 13(1): 1–29.

———. 2007. Bravas, Permitidas, Obsoletas: Mapuche Women in the Chilean Print Media. *Gender and Society* 21(4): 553–578.

———. 2010. Of Indians and Terrorists: How the State and Local Elites Construct the Mapuche in Neoliberal Multicultural Chile. *Journal of Latin American Studies* 42(1): 59–90.

Richards, Paul. 1993. "Cultivation: Knowledge or Performance?" In *An Anthropological Critique of Development*, M. Hobart, Ed. Pp. 61–78. London: Routledge.

Richardson, Laurel. 2000. "Writing: A Method of Inquiry." In *The Handbook of Qualitative Research, 2nd Edition*, N.K. Denzin and Y.S. Lincoln, Eds. Pp. 923–948. Thousand Oaks: SAGE.

Robbins, Paul. 2004. *Political Ecology: A Critical Introduction.* London: Blackwell.

Robinson, William. 2004. *A Theory of Global Capitalism: Transnational Production, Transnational Capitalists, and the Transnational State.* Baltimore: Johns Hopkins Press.

Rocheleau, Dianne. 1995. Maps, Numbers, Texts and Contexts: Mixing Methods in Feminist Political Ecology. *Professional Geographer* 47(4): 458–467.

Rocheleau, Dianne, and D. Edmunds. 1997. Women, Men and Trees: Gender, Power and Property in Forest and Agrarian Landscapes. *World Development* 25(8): 1351–1371.

Rocheleau, Dianne, B. Thomas-Slayter, and E. Wangari. 1996. *Feminist Political Ecology: Global Perspectives and Local Experiences.* London: Routledge.

Rocheleau, Dianne, and Roth. 2007. Rooted Networks, Relational Webs and Powers of Connection: Rethinking Human and Political Geographies. *Geoforum* 38(3): 433–437.

Rohter, Larry. 2004. Mapuche Indians in Chile Struggle to Take Back Forests. *New York Times*, August 11.

Rose, Nikolas. 1999. *Governing the Soul: The Shaping of the Private Self.* London: Free Association Books.

Roué, Maria, and Douglas Nakashima. 2002. Knowledge and Foresight: The Predictive Capacity of Traditional Knowledge Applied to Environmental Assessment. *International Social Science Journal* 54(173): 337–347.

Rutherford, Paul. 1994. The Administration of Life: Ecological Discourse as 'Intellectual Machinery of Government.' *Australian Journal of Communication* 21(3): 40–55.

Saavedra Peláez, Alejandro. 2000. *Los Mapuche En La Sociedad Chilena Actual.* Santiago: LOM Ediciones.
Satterfield, Terre. 2002. *Anatomy of a Conflict: Identity, Knowledge and Emotion in Old-Growth Forests.* Vancouver: University of British Columbia Press.
Sayer, A. 2003. (De)commodification, Consumer Culture, and Moral Economy. *Environment and Planning D* 21: 341–357.
Schmink, Marianne, and Charles Wood. 1987. "The Political Ecology of Amazonia." In *Lands at Risk in the Third World*, P. Little and M. Horowitz, Eds. Pp. 38–57. Boulder: Westview Press.
Schroeder, Richard. 1999. *Shady Politics: Agroforestry and Gender Politics in the Gambia.* Berkeley: University of California Press.
Scott, James. 1975. *Weapons of the Weak: Everyday Forms of Peasant Resistance.* New Haven: Yale University Press.
———.1998. *Seeing Like a State: How Certain Schemes to Improve Have Failed.* New Haven: Yale University Press.
Sedjo, R., and D. Botkin. 1997. Forest Plantations to Spare Natural Forests. *Environment* 39(10): 14–20.
Sedjo, R. A, and S. K. Swallow. 2002. Voluntary Eco-labeling and the Price Premium. *Land Economics* 78: 272–28.
Seguel, Alfredo. 2001. *Invasión Forestal Y Etnocídio Mapuche.* Electronic document, www.wrm.org.uy/paises/Chile/Aseguel.doc, accessed February 5, 2008.
Sen, Sankar, and C.B. Battacharya. 2001. Does Doing Good Always Lead to Doing Better? Consumer Reactions to Corporate Social Responsibility. *Journal of Market Research* 38(2): 225–243.
Sheldon, Christopher. 1997. *ISO 14001 and Beyond: Environmental Management Systems in the Real World.* Greenleaf Publications, 1997.
Sheridan, Thomas. 1988. *Where the Dove Calls: The Political Ecology of a Peasant Community in Northwest Mexico.* Tucson: University of Arizona Press.
Shiva, Vandana. 1988. *Staying Alive: Women, Ecology, and Development.* London: Zed Books.
Sierra, Malú. 1992. *Mapuche, Gente De La Tierra.* Santiago: Editorial Persona.
Silk, John. 1998. Caring at a Distance. *Ethics, Place and Environment* 1(2): 165–182.
Sillitoe, Paul. 1998. The Development of Indigenous Knowledge. *Current Anthropology* 39(2): 223–252.
Sivaramakrishnan, K. 1999. *Modern Forests: Statemaking and Environmental Change in Colonial Eastern India.* Stanford: Stanford University Press.
Sivaramakrishnan, K. 2000. State Sciences and Development Histories: Encoding Local Forestry Knowledge in Bengal. *Development and Change* 31(1): 61–89.
Smith, Geri. 2005. China and Chile: South America Is Watching. *Business Week Online* November 18. Electronic document, www.businessweek.com/bwdaily/dnflash/nov2005/nf20051118_8302_db016.htm?campaign_id=rss_daily, accessed April 11, 2008.
Smith, Neil. 1984. *Uneven Development: Nature, Capital, and the Production of Space.* Oxford: Blackwell.
Star, Susan Leigh, and James Griesemer. 1989. Institutional Ecology, Translations, and Boundary Objects: Amateurs and Professionals in Berkeley's Museum of Vertebrate Zoology. *Social Studies of Science* 19: 387–420.
Starn, Orin. 2011. Here Come the Anthros (Again): The Strange Marriage of Anthropology and Native America. *Cultural Anthropology* 26(2): 179–204.
Starrs, Paul, Carlin Starrs, Genoa Starrs, and L. Huntsinger. 2001. Fieldwork . . . with Family. *The Geographical Review* 91(1–2): 74–87.
Stavenhagen, Rodolfo. 2003. Report of the Special Rapporteur on the Situation of Human Rights and Fundamental Freedoms of Indigenous People: Mission to Chile. UN Economic and Social Council.

Stengers, Isabelle. 2005. The Cosmopolitical Proposal. In *Making Things Public*, Bruno Latour & Peter Weibel, Eds. Pp. 994–1003. Mit Press.
Stengers, I. 2010. *Cosmopolitics*. Minneapolis: University of Minnesota Press.
Stephens, Lynn. 2002. *Zapata Lives! Histories and Cultural Politics in Southern Mexico*. Berkeley: University of California Press.
Stoler, Ann Laura, and Frederick Cooper. 1997. "Between Metropole and Colony: Rethinking a Research Agenda." In *Tensions of Empire*, Cooper and Stoler, Eds. Pp. 1–40. Berkeley, CA: University of California Press.
Strathern, Marilyn, Ed. 2000. *Audit Culture: Anthropological Studies in Accountability, Ethics, and the Academy*. London: Routledge.
Subramaniam, Mangala. 2000. Whose Interests? Gender Issues and Wood-Fired Cooking Stoves. *American Behavioral Scientist* 43(4): 707–728.
Suchman, Lucy. 2007. *Human-Machine Reconfigurations: Plans and Situated Actions*. Cambridge: Cambridge University Press.
Swyngedouw, Erik. 2000. Authoritarian Violence, Power, and the Politics of Rescaling. *Environment and Planning D: Society and Space* 18: 63–76.
Szerszynski, Bronislaw, and John Urry. 2006. Visuality, Mobility and the Cosmopolitan: Inhabiting the World from Afar." *The British Journal of Sociology* 57(1): 113–131.
Taussig, Michael. 1980. *The Devil and Commodity Fetishism in South America*. Chapel Hill: University of North Carolina Press.
———. 1992. *The Nervous System*. Routledge: New York.
Taylor, Peter J., and Frederick Buttel. 1992. How Do We Know We Have Global Environmental Problems? Science and the Globalization of Environmental Discourse. *Geoforum* 23(3): 405–416.
Taylor, P. L. 2005. In the Market But Not of It: Fair Trade Coffee and Forest Stewardship Council Certification as Market-Based Social Change. *World Development* 33(1): 129–147.
Tenner, Edward. 1996. *Why Things Bite Back: Technology and the Revenge of Unintended Consequences*. New York: Knopf.
Timmermans, Stefan, and Marc Berg. 1997. Standardization in Action: Achieving Local Universality through Medical Protocols. *Social Studies of Science* 27: 273–305.
Timmermans, Stefan, and Steven Epstein. 2010. A World of Standards But Not a Standard World: Toward a Sociology of Standards and Standardization*. *Annual Review of Sociology* 36: 69–89.
Tobar, Hector. 2003. When Forests Are Foes. *Los Angeles Times* March 12.
Tockman, Jason. 2004. Anatomy of a Raw Deal in Chile. *Earth First! Journal* 24(3): 12–14.
Toledo, Victor Llancaqeo. 1997. Todas Las Aguas: El Subsuelo, Las Riberas, Las Tierras. *Liwen* 4: 36–79.
Tollefson, Chris. 2004. "Indigenous Rights and Forest Certification in British Columbia." In *Hard Choices, Soft Law*. J. Kirton and M. Trebilcock, Ed. Burlington: Ashgate.
Tsing, Anna L. 2005. *Friction: An Ethnography of Global Connection*. Princeton: Princeton University Press.
Turnbull, David. 2000. *Masons, Cartographers and Tricksters: Comparative Studies of Scientific and Local Knowledge*. Amsterdam: Harwood Academic Publishers.
UNECE (United Nations Economic Commission for Europe). 2007. *Forest Products Annual Market Review*. New York and Geneva: United Nations.
Upton, Christopher, and Stephen Bass. 1996. *The Forest Certification Handbook*. Delray Beach: St. Lucie Press.
Vanclay, J.K., and J.D. Nichols. 2005. What Would a Global Forest Convention Mean for Tropical Forests and for Timber Consumers? *Journal of Forestry* 103(3): 120–125.

Vandergeest, Peter, and Nancy Lee Peluso. 2006. Empires of Forestry: Professional Forestry and State Power in Southeast Asia, Part 1. *Environment and History*: 31–64.

Vayda, Andrew, and Brad Walters. 1999. Against Political Ecology. *Human Ecology* 27(1): 167–179.

Verscheure, H., E. Neira, A. Lara, C. Echeverria, P. Rutherford, and C. Revenga. 2002. *Bosques Frontera De Chile: Un Patrimonio Natural A Conservar*. Santiago: Instituto De Recursos Mundiales.

Viveiros De Castro, Eduardo. 2005. "Perspectivism and Multi-naturalism in Indigenous America." In *The Land Within: Indigenous Territory and the Perception of the Environment*, A. Surrales and P. Garcia, Eds. Pp. 36–74. Copenhagen, Denmark: International Working Group on Indigenous Affairs.

Wallerstein, I. 1974. The Modern World-System I: Capitalist Agriculture and the Origins of the European World-Economy in the Sixteenth Century (Studies in Social Discontinuity). New York: Academic.

Wallerstein, Immanuel. 1979. *The Capitalist World-System*. Cambridge: Cambridge University Press.

Warren, D., et al. 1995. *The Cultural Dimension of Development: Indigenous Knowledge Systems*. London: Intermediate Technology Publications.

Watson-Verran, Helena, and D. Turnbull. 2001. "Science and Other Indigenous Knowledge Systems." In *Handbook of Science and Technology Studies*, S. Jasanoff et al., Eds. London: SAGE.

West, Paige. 2005. Translation, Value and Space: Theorizing an Ethnographic and Engaged Environmental Anthropology. *American Anthropologist* 107(4): 632–642.

———. 2010. Making the Market: Specialty Coffee, Generational Pitches, and Papua New Guinea. *Antipode* 42(3): 690–718.

Whatmore, Sarah, and L. Thorne. 1997. "Nourishing Networks: Alternative Geographies of Food." In *Globalising Food: Agrarian Questions and Global Restructuring*, D. Goodman and M. Watts, Eds. Pp. 287–304. London: Routledge.

White, Andy, Xiufang Sun, Kerstin Canby, Jintao Xu, Christopher Barr, Eugenia Katsigris, Gary Bull, Christian Cossalter, and Sten Nilson. 2006. *China and the Global Market for Forest Products: Transforming Trade to Benefit Forests and Livelihoods*. Washington, DC: Forest Trends.

White, George, and Darius Sarshar. 2004. *Responsible Purchasing of Forest Products*. Gland: WWF-International.

Wiersum, K. Freerk. 1995. 200 Years of Sustainability in Forestry: Lessons from History. *Environmental Management* 19(3): 321–329.

Wilcox, Ken. 1996. *Chile's Ancient Forest: A Conservation Legacy*. Redway, CA: Ancient Forests International.

Wilk, Richard. 2001. Consuming Morality. *Journal of Consumer Culture* 1(2): 245–260.

Williams, Raymond. 1980. *Problems of Materialism and Culture*. London: Verso.

Winner, Langdon. 1999. "Do Artifacts Have Politics?" In *The Social Shaping Of Technology, 2nd Edition*, MacKenzie and Wajcman, Eds. New York: McGraw.

Witte, Ben. 2008. Police Killing of Mapuche Youth Sparks Unrest. *Santiago Times*, January 7. Electronic document, www.santiagotimes.cl/santiagotimes/2008010612635/news/political-news/police-killing-of-chile-mapuche-youth-sparks-unrest.html, accessed April 10, 2008.

Wolf, Eric. 1982. *Europe and the People without History*. Berkeley: University of California Press.

Wolodarsky-Franke, Alexia, and Antonio Lara. 2005. The Role of "Forensic" Dendrochronology in the Conservation of Alerce (Fitzroya cupressoides ((Molina) Johnston)) Forests in Chile. *Dendrochronologia* 22(3): 235–240.

Yearley, Stephen. 2001. "The Environmental Challenge to Science Studies." In *Handbook of Science and Technology Studies*. S Jasanoff et al., Eds. Pp. 457–489. London: SAGE.

Index

agrarian reform 9, 44, 50, 115
Agrawal, Arun 5, 15, 30, 71, 91, 125
agriculture 3, 8, 12, 15, 29–30, 68, 101–3, 116
AIFBN (Agrupación de Ingenieros Forestales por el Bosque Nativo) 1, 40–1, 89–91, 109
alerce 110–11, 121n7
anthropology 5, 25n4, 60, 76, 80n3; and indigenous knowledge 60–2, 70–1; and science studies 14, 23, 25n9
Astorga, Luis 1, 38, 40–1, 63–6, 89, 99, 109, 114
auditing 2–3, 18, 28, 40–1, 57, 66–7, 108–12
Auracanía 9, 49–50, 54n9, 67
auracaria 8, 110
authority 1, 5, 16, 29–30, 59–60, 66–7, 75, 112, 112–16, 127

Berlin, Brent 60, 76
biodiversity 1, 6, 8, 31, 58–9, 69, 79, 81n6, 117, 122–3
bosque nativo see native forest
boundary work 9, 16–17, 24, 31–2, 112, 120, 129

campesinos 21, 23, 50, 68, 80n1, 96–7, 103, 115, 123–4
certification ix–x, 1–5, 11–13, 27–32, 34–5, 39–41, 43–6, 53–4, 56–9, 65–6, 69, 80n2, 83, 89–93, 96, 99, 104–8, 112–13, 118–20, 125–32
chain of custody 28, 41, 57, 90, 104–8
COCEL (Consejo de Certificación de Leña) 89, 91

CODEFF 40, 46, 102–3, 109,
colonialism 7, 43–4, 49, 70–1, 75, 7–80, 81n5, 115
commodities 6, 11–13, 28, 33, 35–6, 82, 123–6
commodity fetish 11–13, 33, 123–6
CONAF (Comisión Nacional Forestal) 65, 73, 90, 104–6, 113–14, 121n5
Concepcíon 65
consumer 1–3, 6, 12–13, 28, 32–6, 39–40, 57, 82–3, 90–3, 107–8, 124–5, 127–8
CORMA (Corporacion Nacional de la Madera) 35, 43
cosmopolitics 3, 14, 25n9, 76–7, 80, 130
Cowan, Ruth Schwartz 24, 128–30

dictatorship x, xin1, 9, 11–12, 25n6, 38–9, 48–9, 50–1, 115
Dove, Michael 78–9
Dunn, Elizabeth 29–30, 120

embodiment 12–13, 61–2, 68, 82–3, 93–4, 125, 130
environmentalism 7–9, 17, 19–22, 36–41, 43, 45, 48, 52–3, 58–9, 117–18, 126
ethical trade 6, 12, 33–6, 93
ethics 1, 6–7, 13, 14, 40, 93–4, 125–6, 129–33
ethnobiology 60–1, 70, 74–6
ethnography 5, 7–8, 17–22, 33–4, 50, 82, 86

Fair Trade 1, 6, 12–14, 25n7, 28, 35, 93, 131

154 *Index*

firewood 8, 10, 57, 82–94, 104
forest 3, 5, 7–11, 13–15, 18–19, 20–2, 27, 29, 35, 39, 44–5, 46–7, 52, 69, 72–4, 90–1, 102, 109–11, 122
Forest Stewardship Council (FSC) 1–3, 6, 14, 28, 32, 34, 39, 43, 51–2, 56–9, 62, 112; FSC-Chile 1–3, 15–16, 17–18, 35, 40, 41, 44–5, 47, 53, 60, 63–6, 89, 95, 99, 100, 109, 112, 114, 121n5, 130–2; FSC-International 31, 34, 40–1, 54n2, 108, 113
forestales see timber companies
forestry 1–4, 8–9, 28, 30–2, 41–7, 51, 56–9, 71–4, 77, 80n4, 91, 100–4, 105–8
Foucault, Michel 5–7, 14, 24, 25n10, 30, 78, 126

gender 16, 81n5, 97–8, 118, 128
global warming 44–5, 54n7, 118, 125
Guthman, Julie 12, 25n3, 28, 34, 116

Haraway, Donna 14, 21, 29, 31, 93–4, 130
health 10, 33, 62–3, 65–6, 68, 83, 86, 90–4
Huaiquilao, Pablo 19, 53–4, 68–9, 71–5, 80

ICEFI (Initiativa pro Certificacion Forestal Independiente) 17–19, 34, 41
indigenous peoples 3, 6, 11, 14, 22, 27, 31, 49, 51, 54n2, 61, 69–71, 75–6
INFOR (Instituto Forestal) 42–4
Ingold, Timothy 61, 74, 117

knowledge ix–x, 1, 5, 13–16, 60, 75–80, 81n5, 94, 117, 122, 129–30; indigenous 51–2, 60–2, 67–75

labor 11–13, 19, 24, 28, 31, 64–6, 80, 100, 103–4, 115, 128–9
land 8–10, 44, 48, 49, 50–1, 96–7, 115–16
Latour, Bruno 12, 14–16, 27–8, 32, 70, 76, 108, 119, 127
LEED 40
Leña see firewood
Lowe, Celia 79–80, 116

Mapuche 9–11, 17, 21–2, 44, 47–54, 66, 67–9, 71–5, 111, 114–15
Marx, Karl 11–13, 82, 93, 123, 131
Masisa 43, 45, 105–7
Mawidakom 46–7

native forest 7–10, 44, 58, 72–4, 80, 83, 90–1, 101–2, 106, 112–13, 115
nature ix–x, 1, 15, 20, 29–32, 60–2, 69, 72–4, 79, 117–19
neoliberalism 1–3, 6, 11, 39, 44, 59–60, 114–15, 120, 130–2
network 12–16, 28, 88, 92, 104–8, 125

organic 3, 12–13, 25n3, 28, 33, 35, 54n3, 93, 116
Otero, Luis 38, 40–1, 54n6, 100–2, 108–12

Paglia, Todd 36
peasants *see campesinos*
pine 9–11, 44, 72, 85, 88, 102, 109–11, 114–14
Pinochet, Augusto xi, 10–11, 38–9, 49, 51, 114–15
plantations 8–11, 18, 34, 42, 44, 49, 54n7, 58, 72–5, 78, 80n4, 102, 109, 114–14
policy 4, 16, 19, 25n13, 27, 64–5, 71, 78, 112–16, 125, 129
political ecology 5, 38, 44, 96–9, 132
postcolonial 7, 75, 77–9, 80, 117
Principles, FSC 2–3, 31–2, 57–9, 68, 100, 108, 115; Principle 3 11, 51–4, 100; Principle 4 58, 62–6, 100

reason x, 75–80, 81n6, 118, 132
resistance 12, 49–50, 67, 120, 124, 133
responsibility 5, 31–2, 43, 45, 57–60, 66, 90, 94, 113, 125, 130–3
scale 17, 30–1, 34, 46–7, 58, 82–3, 92–4, 95–120, 125, 127
science studies 5, 14, 16–17, 23, 25n9, 127–8
Sivaramakrishnan, K. 77–8
small proprietors 19, 24, 46–7, 65, 80, 100–4, 120, 131
stakeholders xi, 3, 5, 27, 31–2, 37, 52–4, 65, 71, 80, 93, 123, 130

standards 1–4, 11, 12–13, 14–17, 23, 28, 29–30, 39, 41, 43, 45–6, 52–3, 54n8, 56, 60, 62–3, 65–7, 75, 80, 80n2, 91–2, 100, 108–12, 119–20, 125
state, the x, 3–4, 25n13, 50, 53, 104, 112–16, 120, 127
subjectivity 5–6, 7, 15, 27–32, 54, 59–60, 94, 100, 125, 129–31
sustainability xi, 1–4, 5–6, 29, 33, 35, 40–1, 45, 57, 59, 93, 118, 125

technologies of value 5–7, 13–14, 24, 27, 30–1, 34, 54, 104, 126–30, 132–3
technology 37, 86, 114, 126–8; appropriate 86
Temuco 50, 65, 67, 72, 89–90, 91

territory 6, 9–11, 44, 48, 50–4, 58, 115
timber companies 1, 3–6, 9–11, 41–44, 52, 53, 56–60, 66, 68–9, 74, 80n4, 105–6, 114, 116
timber industry 8–9, 28, 31, 35, 41, 43–45, 101, 111, 115, 118
Tsing, Anna 17, 27, 104, 107, 117–18, 122

Valdivia ix, 18, 38, 49–50, 56, 65, 84, 88–91
value 1–7, 11–13, 30–1, 93–4, 99, 104, 116, 124, 126–8, 130–3
virtualism 44, 66–7, 130–2

wood ix, 1–3, 6, 8, 11–13, 28, 33–5, 40, 41–3, 56–8, 69, 82–3, 102, 105–8, 125, 131